Engineering Seismology, Geotechnical and Structural Earthquake Engineering

Engineering Seismology, Geotechnical and Structural Earthquake Engineering

Editor

Hubert O'Kelly

scitus
academics

Engineering Seismology, Geotechnical and Structural Earthquake Engineering
Edited by **Hubert O'Kelly**

ISBN: 978-1-68117-303-0
Library of Congress Control Number: 2016937488

www.scitusacademics.com
Box No. 4766, 616 Corporate Way,
Suite 2, Valley Cottage,
NY 10989

Printed in United States of America on Acid Free Paper

Preface

Seismology is the study of seismic waves. It is the scientific study of earthquakes and the spread of elastic waves across the Earth or other planet-like bodies. Studies of the environmental effects of earthquakes—tsunamis and various seismic sources such as volcanic, tectonic, oceanic, atmospheric and artificial methods, are also included. An earthquake is the energy that moves through the earth after an earthquake. Some earthquake waves can be felt on the earth's surface and others can't. In their work, seismologists use special tools called seismographs. These machines have recorded seismic waves. They can usually detect and amplify the slightest movement on the ground. Two types of shock waves are typically used for seismology studies. Push waves occur when the inner structure of the earth moves forward. Then there are torsion waves that lead to twisted particles between sliding inner structures. The time and speed of these waves can give seismologists valuable information like the distance or severity of an earthquake. Seismology also credits people with an accurate base of knowledge about the inner layers of the earth. n

Earthquake–hazard mitigation in modern society plays an important role. Mitigation of such hazards involves detailed seismicity studies, site effects and seismic landslides, as well as structural design and analysis to withstand such actions. The earthquake study combines science, technology and infrastructure with engineering expertise to reduce human and material losses when they inevitably occur.

The book deals with seismic hazard analysis, seismic design, assessment and remediation, earthquake physics and soil structure interaction analysis.

Table of Contents

1

CHAPTER

SEISMIC HAZARD ANALYSIS FOR ARCHAEOLOGICAL STRUCTURES — A CASE STUDY FOR EL SAKAKINI PALACE CAIRO, EGYPT

Sayed Hemeda[1]

[1] Conservation Department, Faculty of Archaeology, Cairo University, Egypt

1. INTRODUCTION

The modern architectural heritage of Egypt is rich, and extensively variable. It covers all kinds of monumental structures from palaces, public buildings, residential and industrial buildings, to bridges, springs, gardens and any other modern structure, which falls within the definition of a monument and belongs to the Egyptian cultural heritage. We present herein a comprhensive geophysical survey and seismic hazard assesment for the rehabilitation and strengthening of Habib Sakakini's Palace in Cairo, which is considered one of the most significant architectural heritage sites in Egypt. The palace located on an ancient water pond at the eastern side of Egyptian gulf close to Sultan Bebris Al-Bondoqdary mosque, a place also called "Prince Qraja al-Turkumany pond". That pond had been filled down by Habib Sakakini at 1892 to construct his famous palace in 1897.

Various survey campaigns have been performed comprising geotechnical and geophysical field and laboratory tests, aiming to define the physical, mechanical and dynamic properties of the building and the soil materials of the site where the palace is founded. All these results together with the seismic hazard analysis will be used for the seismic analysis of the palace response in the framework of the rehabilitation and strengthening works foreseen in a second stage. We present herein the most important results of the field campaign and the definition of the design input motion.

The seismic hazard analysis for El Sakakini Palace has been performed based on historical earthquakes, and maximum intensity.PGA with 10% probability of exceedance in 50 and 100 years is found equal to 0.15g and 0.19g respectively. P-wave and S-wave seismic refraction indicated a rather low velocity soil above the seismic bedrock found at depths higher than 20m. Ambient noise measurements have been used to determine the natural vibration frequency of soil and structure of El-Sakakini Palace. The fundamental frequency of El-Sakakini palace is 3.0Hz very close to the fundamental frequency of the underlying soil, which makes the resonance effect highly prominent.

Some floors are considered dangerous since it show several resonance peaks and high amplification factors (4th and 5th floors) these floors are made of wood so, warnings to decision makers are given for the importance of such valuable structures.

The seismic design and risk assessment of El Sakakini palace is performed in two steps. In the first one we perform all necessary geotechnical and geophysical investigation together with seismic surveys and seismic hazard analysis in order to evaluate the foundation soil properties, the fundamental frequency of the site and the structure, and to determine the design input motion according to Egyptian regulations. The second phase comprises the detailed analysis of the palace and the design of the necessary remediation measures. IN the present pare we present the results of the first phase.

2. SEISMIC HAZARD

2.1. Historical seismicity

Egypt possesses a rich earthquake catalogue that goes back to the ancient Egyptian times. Some earthquakes are reported almost 4000 years ago. Figure 1 shows the most important historical events affecting ElSakakini palace. We can see that the Faiyum area as well as the Gulf of Suez is the most important earthquake zones affecting the place.

2.2. Maximum intensity

Historical seismicity and maximum reported intensity is a good preliminary index of the expected severity of a damaging earthquake. Available isoseismal maps in the time period 2200 B.C. to 1995 were digitized and recontoured to determine the maximum intensity affecting the place. This was done using a cells value of equal area 0.1 lat. × 0.1 long. Figure 2 present the produced IMM intensity showing that a maximum IMM of VII is good design value.

2.3. Probabilistic hazard assessment

An improved earthquake catalogue for Egypt and surrounding areas affecting El Sakakini Palace has been prepared for the purposed of this study partially based on recent work of Gamal and Noufal, 2006. The catalogue is using the following sources:

- For the period 2200 B.C to1900: Maamoun,1979; Maamoun et al., 1984 ; Ben-Menahem 1979 and Woodward-Clyde consultants, 1985.
- For the period 1900 to 2006: Makropoulos and Burton, 1981; Maamoun et al., 1984 ; Ben-Menahem 1979; Woodward-Clyde consultants, 1985; Riad and Meyers, 1985; Shapira, 1994 and NEIC, 2006; Jordan seismological observatory 1998-2000.

The horizontal peak ground acceleration over the bedrock of El Sakakini area was estimated using Mcguire program 1993. 37 seismic source zones were used to determine the horizontal PGA over the bedrock (Figure 3), while PGA attenuation formula of Joyner and Boore, 1981 was used because of its good fitting to real earthquake data in Egypt. A complete analysis for the input parameters to estimate the PGA values over the bedrock can be found in Gamal and Noufal, 2006.

$$(PGA) = 2.14\, e1.13^{M} D^{-1} e^{-0.00590} \quad D = \left(R^2 + 4.0^2\right)^{0.5} \tag{1}$$

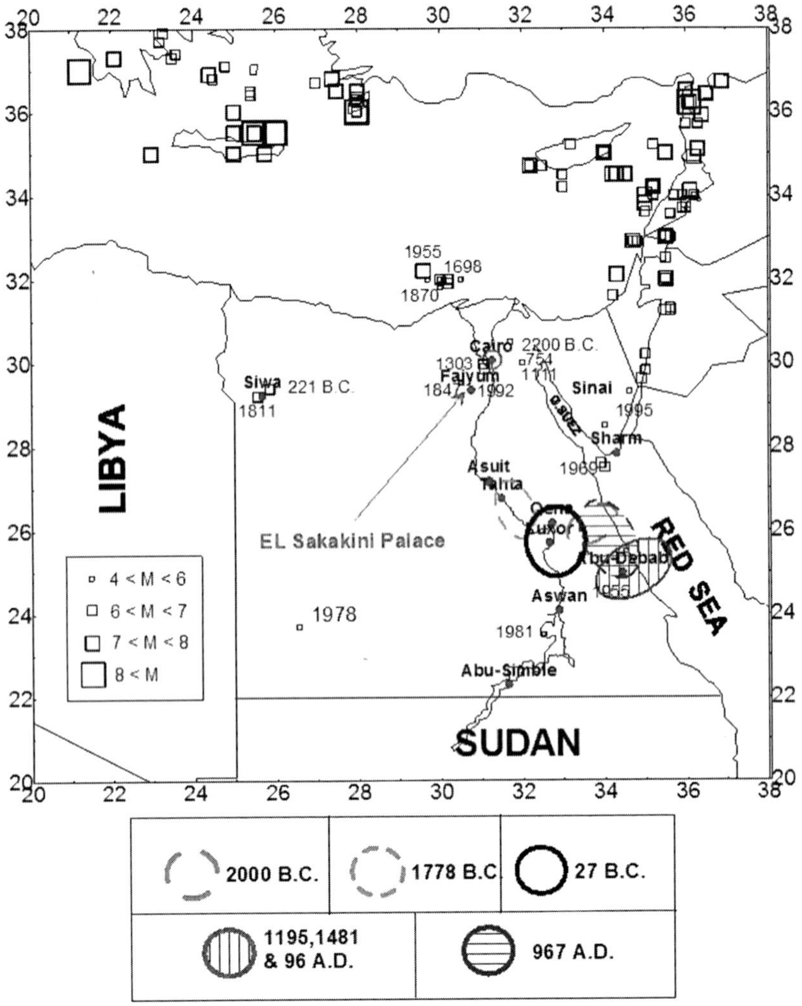

Figure 1. Important and historical earthquakes occurred in and around El Sakakini Palace area in the period 2200 B.C to1995.

The probabilistic analysis provided the following results: The peak horizontal acceleration in gals with 10 % probability of exceedance over 50 years is 144cm/sec^2(or 0.147g) For 10% probability in 100 years the estimated PGA for rock conditions is 186 (cm/sec^2) (or 0.19g)(Figures 3 and 4). These values are quite high and considering the local amplification they may affect seriously the seismic design and stability of El. Sakakini Palace.

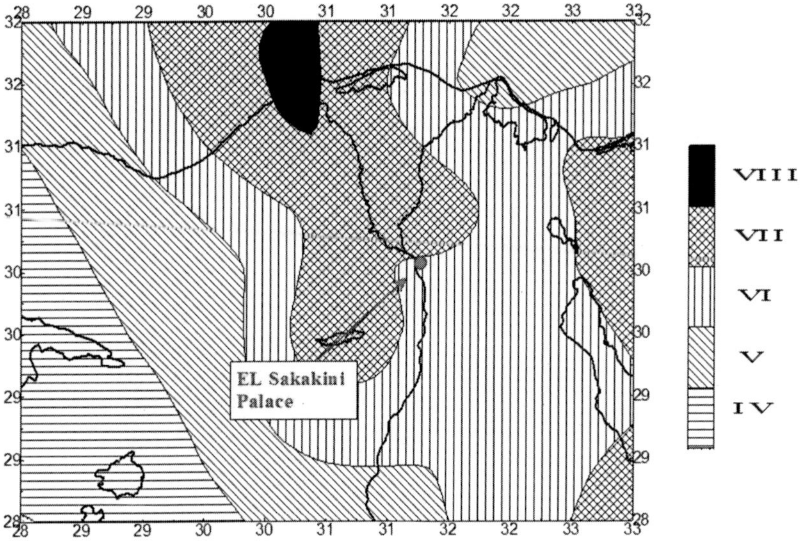

Figure 2. Maximum intensity zonation map based on the historical seismicity reported in the time period 2200 BC to1995

3. GEOTECHNICAL INVESTIGATION

Core drilling is among the routine methods for subsurface exploration. Most commonly, NX-size core drill is used, representing a hole diameter of 76 mm (3") and a core diameter of 54 mm (2 1/8"). The drilling often has multiplier purposes, of which the following are in most cases the most important:

Verification of the geological interpretation. Detailed engineering geological description of rock strata. To obtain more information on rock type boundaries and degree of weathering. To supplement information on orientation and character of weakness zones. To provide samples for laboratory analyses. Hydro geological and geophysical testing. Input data for engineering classification of rock masses.

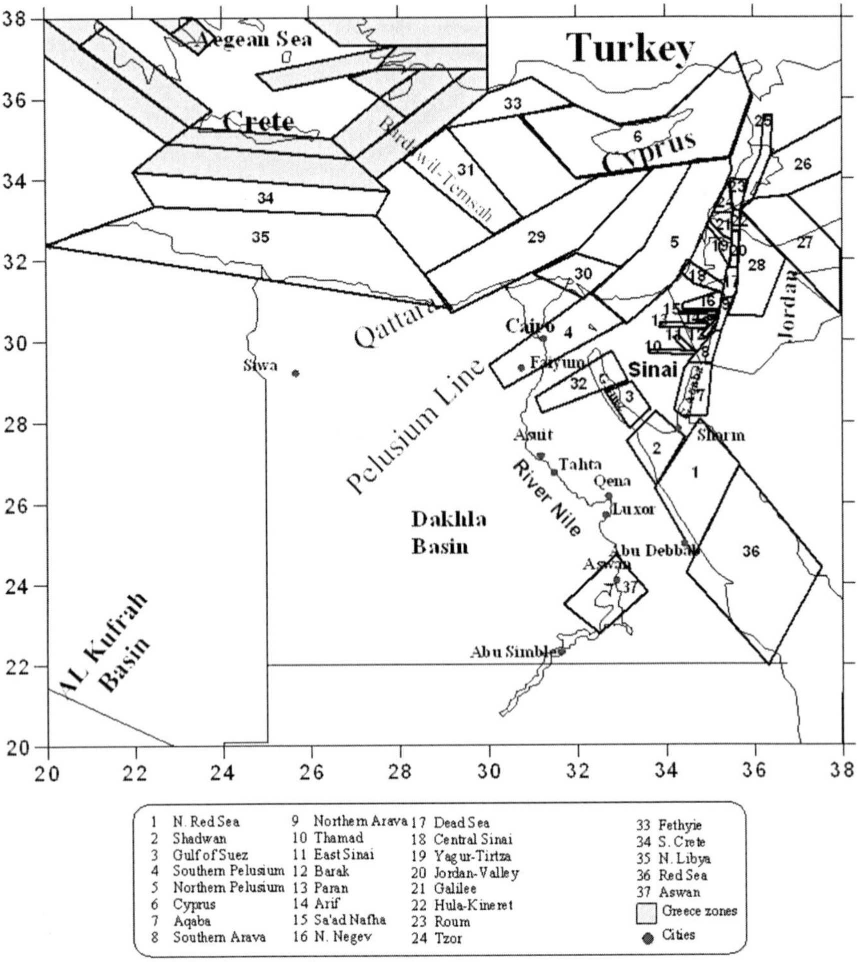

Figure 3. Seismic source regionalization using 37 seismic source zone (except greece zones) adopted for Egypt and surrounding areas (Gamal and Noufal, 2006).

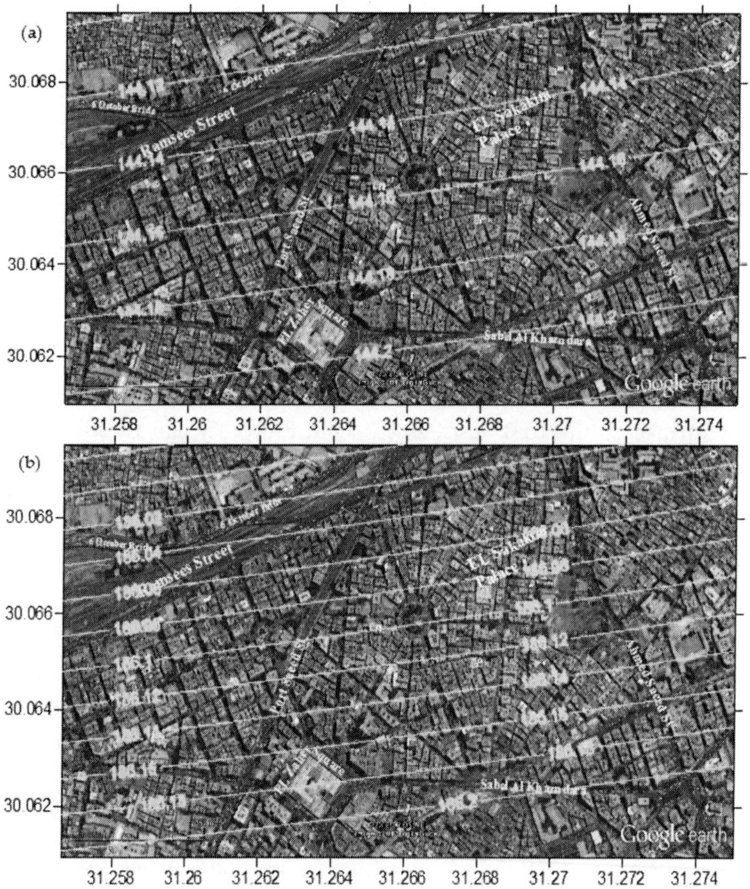

Figure 4. a: Peak Horizontal Acceleration in gals (cm/sec^2) for the seismic bedrock with10 % probability of exceedance in 50 years. b: Peak Horizontal Acceleration in gals (cm/sec^2) for the seismic bedrock with 10 % probability of exceedance over 100years

The geotechnical investigation, six geotechnical boreholes with Standard Penetration Test (SPT) measurements have been carried out in the archaeological site included the drilling of three geotechnical boreholes with integral sampling to a depth 20 meters, one borehole to depth 15 meters and two boreholes to depth 10 meters at six locations in the site. The geotechnical data also indicated the ground water level at the archaeological site. We did all the boreholes inside the site with hand boring machine.

The results of laboratory tests which have been carried out on the extracted soil samples from the boreholes, which include specific gravity (Gs), water content (Wn), saturated unit weight (γsat), unsaturated unit weight (γunsat), Atterberg limits and uniaxial compressive strength (UCS), in addition to the ground water table (GWT), are shown in the figures (7a,7b).

The shear wave profile obtained by using ReMi compared very well to geotechnical boreholes and geophysical survey data. In addition, the shear wave profile obtained by using ReMi Performed much better than commonly used surface shear-wave velocity measurements.

Geotechnical boreholes (1) through (3) indicated that:

Filling of Fill (silty clay and limestone fragments, calc, dark brown) From ground surface 0.00m to 3.50m depth. Sand Fill (silty clay, medium, traces of limestone& red brick fragments, calc, dark brown) From 3.50m to 5.00m depth. Silty clay, stiff, calc, dark brown From 5.00m to 6.50m depth. Clayey silt, traces of fine sand & mica, yellowish dark brown From 6.50m to 8.50m depth. Silty sand, fine, traces of clay & mica. Dark brown. From 8.50m to 11.00m depth. Sand, fine, some silt, traces of mica, yellowish dark brown. From 11.00m to 14.00m depth. Sand, fine to medium, traces of silt& mica, tracesof fine to medium gravel, traces of marine shells, yellowish dark brown. From 14.00m to 16.00m depth. Sand, fine, traces of silt & mica, yellowish dark brown. From 16.00m to 18.00m depth. Sand & Gravel, medium sand, graded gravel, traces of silt, yellow darkbrown. From 18.00m to 20.00m depth. End of drilling at 20.00m.

Geotechnical boreholes (4) through (6) indicated that:

Fill (silt, clay and fragments of limestone and crushed brick, from ground surface 0.00m to 4m depth. Fill (silty cal with medium pottery and brick fragments, calc dark brown) from 4 to 5 m depth. Brown stiff silty clay and traces of limestone gravels, from 5.00m to 7.50m depth. silt, traces of brown fine sand & traces of clay from 12.00m to 14.00m depth. Dark brown clay silt with traces of fine sand. from 14.00m to 15.00m depth.

Figure 5. El- Sakakini palace and the Geotechnical investigations.

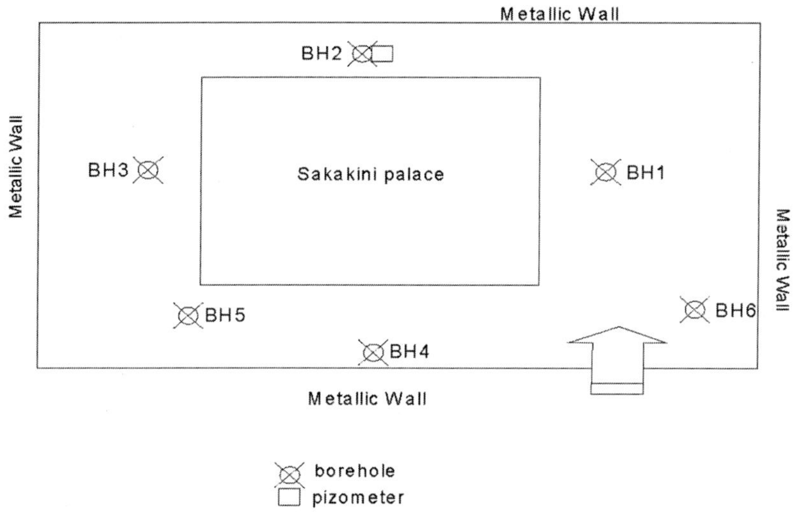

Figure 6. General layout & boreholes locations.

project : existing . habib pasha elsakakeeny palace	file no : sakakeeny feb10
location :eldaher-cairo	date commenced : jan . 15- 2012
Boring. no : 1	datecompleted : jan . 18- 2012
DRILL METHOD : MANUAL DRILLING	weather : cold
drill fluid :none	ground level :
driller : alaa amin drilling co	initial / final gwd : 2.30/1.20

depth (m)	legend	classification	end of layer (m)	qu (kg/ cm2)	spt n/30 cm	yb (um3)	f.s. (%)	wL (%)	wP (%)	RECOVERY (%)	R.Q.D. (%)
1		fill(silty clay and limestone fragments . calc dark brown									
2											
3				1.10							
4		fill(silty clay .medum u of limestone &red bock fragments .calc dark browen)	3.40	1.00							
5			4.80								
6		silty clay stiff calc dark browen		1.80							
7		clayey silt . traces of fine sand & mica yellowsh dark brown	6.30	1.80							
8			8.00								
9		more sand more silt	8.50	0.80				39.82	21.42		
10		silty sand fine traces of clay &mica dark brown	9.50								
11			11.00								
12		sand fine some silt traces of mica yellowish dark brown			21						
13											
14		sand fine to med u of silty clay dark brown	13.00 14.00								
15		sand fine to medum traces of silt & mica of fine to medum gravel u of marine shells yellowish dark brown									
16			16.00								
17		sand fine traces of silt & mica yellowish dark brown			28						
18			18.00								
19		sand & gravel medum sand graded gravel u of silt yell dark brown									
20			20.00		33						
		end of drilling at 20.00m									

project : existing . habib pasha elsakakeeny palace	file no : sakakeeny feb10
location :eldaher-cairo	date commenced : May . 15- 2012
Boring. no : 4	datecompleted : May . 18- 2012
DRILL METHOD : MANUAL DRILLING	weather : cold
drill fluid :none	ground level :
driller : alaa amin drilling co	initial / final gwd : 1.10 m

depth (m)	legend	classification	end of layer (m)	qu (kg/ cm2)	spt n/30 cm	yb (um3)	f.s. (%)	wL (%)	wP (%)	RECOVERY (%)	R.Q.D. (%)
1		fill(limestone fragments concrete frag.sand&silt calc dark brown									
2											
3		fill(silty clay .tr of limestone &red brick &pottery fragments .calc dark browen)									
4											
5		fill(silty clay .medum u of pottery fragments .calc dark browen)						51 75	23 95		
6											
7			7.40								
8				2.75							
9				2.00							
10				2.25							
11		clayey silt . traces of mica .dark brown		2.20							
12		clayer silty sand . fine . traces of rubble	12.10	2.00				35 76	18 09		
13		dark brown silt clay with traces of fine sand			31						
14			14.00								
15		fine brown silt and sand with traces of clay &	15.00								
		end of drilling at 15.00m									

Figure 7. a. Geotechnical Borehole_1, El Sakakini Palace. b. Geotechnical Borehole_4, El Sakakini Palace.

4. GEOPHYSICAL CAMPAIGN

4.1. P-wave refraction

A total of 10 seismic profiles are conducted at El Sakakini palace area (Figure 8). All profiles are carried out using 12 receivers, P-type geophones with 5m intervals and 2 shots. The forward and reverse shots were carried at a distance of 1 m at both ends. The seismic shots layouts are described in Table 1.

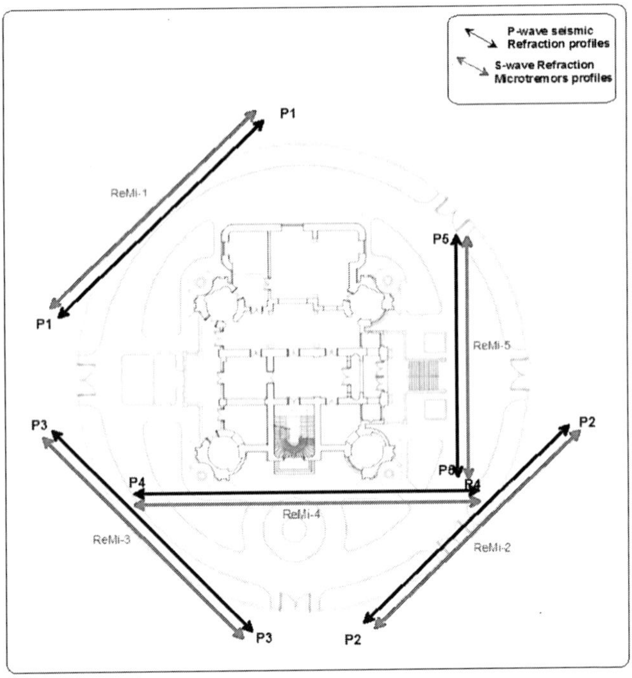

Figure 8. Location of the P-wave seismic refraction, S-wave refraction and ReMiprofiles conducted at ElSakakini Palace.

Table 1. Seismic shots.

Shot #	Name	Offset X (m) (relative to R1)
S2	Forward	-1
S4	Reverse	56

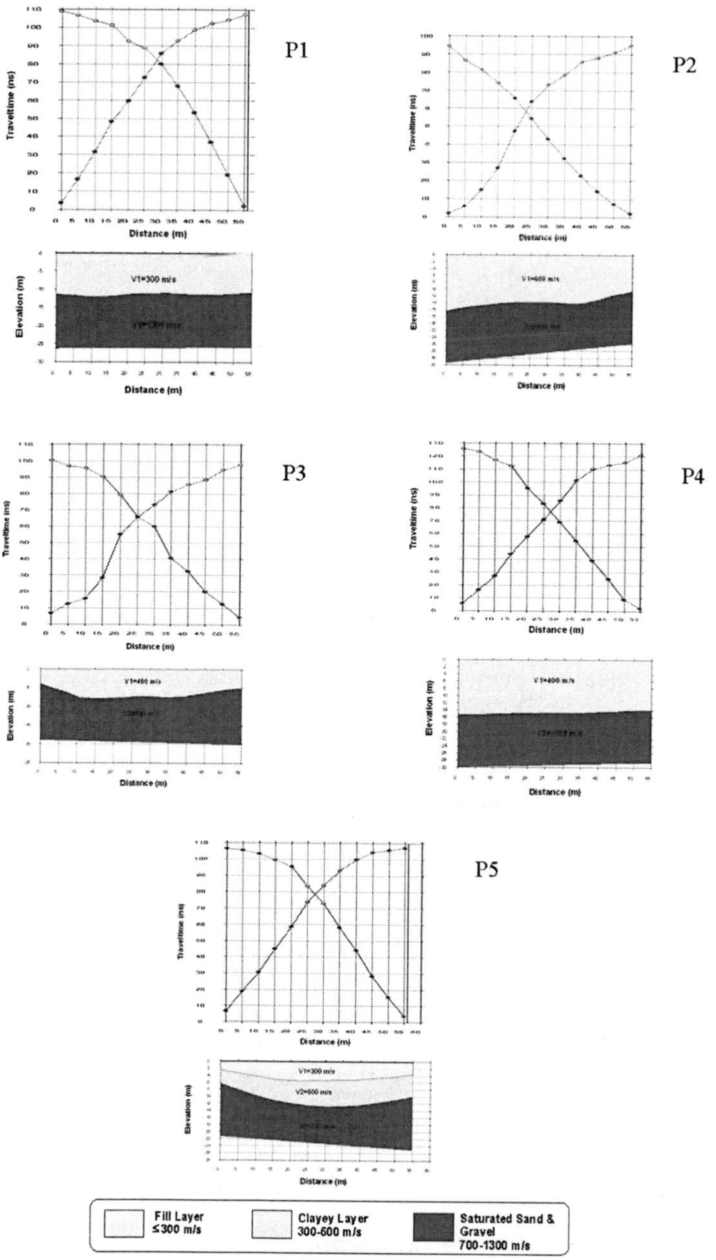

Figure 9. P-wave travel time distance curve and its corresponding geoseismic model for profiles # P1-P5 (Figure 8).

The conducted profiles are interpreted using time-term inversion method; an example of the conducted profiles and corresponding geoseismic model is shown in Figure 9. Table 2 summarizes the measured Vs values and the corresponding soil thicknesses. The soil stratification is not uniform and horizontal, as it should be expected for a filled area. However it is possible to distinguish the following three main layers: the soil layering can be summarized in the following (table 2).

Soil A- Fill (<300 m/s): A surface highly heterogeneous material (mainly man-made fill) with an average thickness of 10 m and an average velocity Vs lower than 300m/s. It is composed of very loose and low strength sediments such as silt, clay and limestone fragments. It is not found in all locations.

Soil B-Clayey soil (400-600 m/s): Below the surface layer (soil A) there is a clayey or silty clay layer with an average thickness of 10 m meters and Vs velocity 400-600 m/s.

Soil C-Saturated Sand & Gravel (700-1300 m/s): Below soil B there is a stiff soil layer with various thicknesses. it shows a considerable increase of Vs seismic velocity reaching sometimes values as high as 1300m/s. The soil is composed of compacted stiff saturated sand and gravel with an average Vs velocity equal or higher than 700m/s. It may be considered as the "seismic bedrock" for the local site amplification analyses.

Table 2. P-wave refraction geophysical campaign conducted at El-Sakakini palace area.

	Layer A	Layer B		Layer C	
Profile N°	Velocity in m/s	Velocity in m/s	Depth (m)	Velocity in m/s	Depth in (m)
1		300	10	1300	25
2		600	16	900	32
3		400	9	700	20
4		400	14	1200	30
5	< 300	300	5	500	10

4.2. Refraction- microtremor (ReMi method)

We have used the ReMi (refraction microtremors) method to determine the S-wave seismic velocity with depth. The method is based on two fundamental ideas. The first is that common seismic-refraction recording equipment, set out in a way almost identical to shallow P-wave refraction surveys, can effectively record surface waves at frequencies as low as 2 Hz (even lower if low frequency phones are used). The second idea is that a simple, two-dimensional slowness-frequency (P-f) transform of a microtremors record can separate Rayleigh waves from other seismic arrivals, and allow recognition of true phase velocity against apparent velocities. Two essential factors that allow exploration equipment to record surface-wave velocity dispersion, with a minimum of field effort, are the use of a single geophone sensor at each channel, rather than a geophone "group array", and the use of a linear spread of 12 or more geophone sensor channels. Single geophones are the most commonly available type, and are typically used for refraction rather than reflection surveying. There are certain advantages of ReMi method: it requires only standard refraction equipment, widely available, there is no need for a triggering source of energy and it works well in a seismically noisy urban setting. (Louie, 2001, Pullammanappallil et al. 2003).

A 12 channel ES-3000 seismograph was used to measure background 'noise' enhanced at quiet sites by inducing background noise with 14Hz geophones in a straight line spacing 5m Figure 5 shows the map were ReMi measurements were made. Almost all the sites were noisy. In particular big hammer used to break some rocks generated noisy background at El Sakakini Palace.30 files of 30sec records (unfiltered) of 'noise' were collected at each site. Five profiles were taken inside the Palace (Figure 8). Figure 10-11 shows an example of the dispersion curves and its P-F image (Remi Spectral ratio of surface waves) for refraction microtremors profile ReMi-1. The estimated average Vs for all profiles are shown in Figure 12.

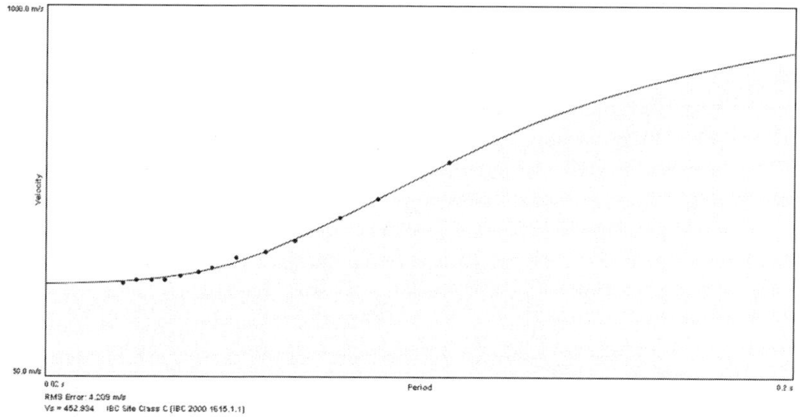

Figure 10. Dispersion curve showing picks and fit for Profile ReMi-1

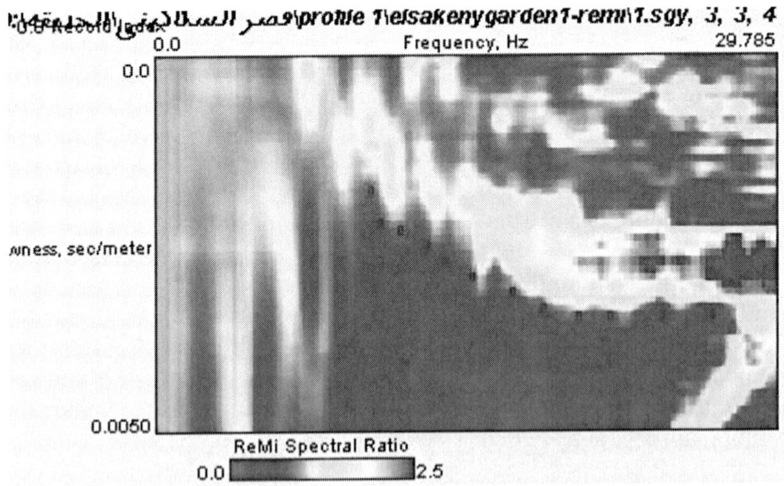

Figure 11. P-F image with dispersion modeling picks for Profile ReMi-1

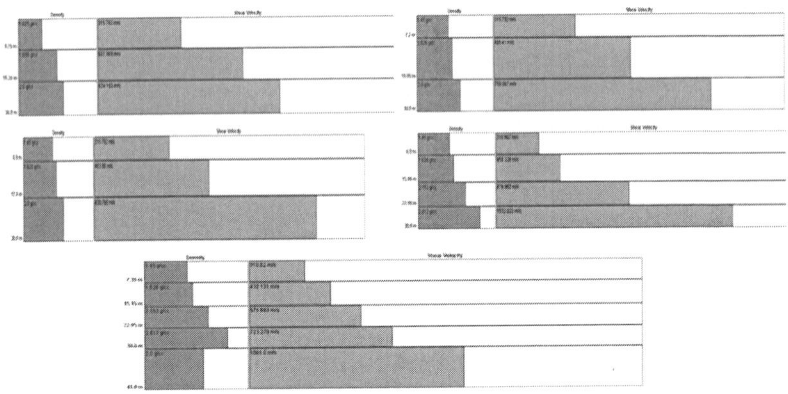

Figure 12. Shear wave velocity model calculated for refraction microtremors profiles ReMi-1 To ReMi-5 (Figure 9).

5. FREQUENCY CHARACTERSITICS OF THE SOIL AND THE BUILDING USING MICROTREMRS

Microtremors are omnipresent low amplitude oscillations (1-10 microns) that arise predominantly from oceanic, atmospheric, and urban or anthropogenic actions and disturbances. The implicit assumption of early studies was that microtremors spectra are flat and broadband before they enter the region of interest (soil or building). When microtremors enter preferable body it changes and resonate depending on the nature of the material, shape, and any other characteristics of this body.

It may be considered to compose of any of seismic wave types. We have two main types of microtremors, Local ambient noise coming from urban actions and disturbances and long period microtremors originated from distances (e.g. oceanic disturbances). There is still a debateongoing on the characteristics of the ambient noise that should be used for site characterization and ground response. While some are using only the longer period microtremors originated from farther distances (e.g. Field et al, 1990), others considered that traffic and other urban noise sources are producing equally reliable results. In general low amplitude noise measurements comparable results give with strong motion data (Raptakis et al, 2005., Pitilakis, 2011., Apostolidis et al., 2004., Manakou et al, 2010., Mucciarelli, 1998).

Kanai 1957, first introduced the use of microtremors, or ambient seismic noise, to estimate the earthquake site response (soil amplification). After that lots of people followed this work but from the point of soil amplification of earthquake energy for different frequencies (e.g. Kanai and Tanaka 1961 and Kanai 1962, Kagami et al, 1982 and 1986; Rogers et al., 1984; Lermo et al., 1988; Celebi et al. 1987).

5.1. Instrumentation and data acquisition

A high dynamic range Seismograph (Geometrics ES-3000 see Figure 13) mobile station with triaxial force balance accelerometer (3 channels), orthogonally oriented was used. The station was used with 4Hz sensors to record the horizontal components in longitudinal and transverse directions in addition to the vertical components. For the data acquisition and processing we followed the following steps:

- Recording 10-min of ambient noise data using a mobile station moving among variable soil stations or El Sakakini building floors/
- Zero correction to the total 10-minnoise at time domain
- Subdivision of each 10-minsignal into fifteen 1-min sub-windows,

Each of these series was tapered with a 3-sec hanning taper and converted to the frequency domain using a Fast Fourier transform,

Figure 13. High dynamic range ES-3000 Geometrics mobile station and triaxial geophone used 4 Hz to drive soil response of El-Sakakini Palace.

- Smoothing the amplitude spectrum by convolution with 0.2-Hz boxcar window,
- Site response spectrum for a given soil site (or certain floor) is given by dividing the average spectrum of this site over the spectrum of the reference site. The reference site is choose carefully in the site as deepest and calmest station in the basement floor with least soil response (usually we choose a certain basement floor location with least soil response to be used as reference site).
- Smoothing the final response curves by running average filter for better viewing. A complete description of the methodology can be found in Gamal and Ghoneim, (2004).

5.2. Ground response

Figure 14 shows the locations of microtremors stations used to determine the ground response at EL Sakakini Palace area. The predominant frequency of the ground at EL Sakakini Palace is about 3 Hz (see Figure 15 & Table 1), a value almost identical to the theoretical estimation according to Kennett and Kerry (1979) (Figure 16 & Table 4). The amplification factor is about 2, which is relatively low.

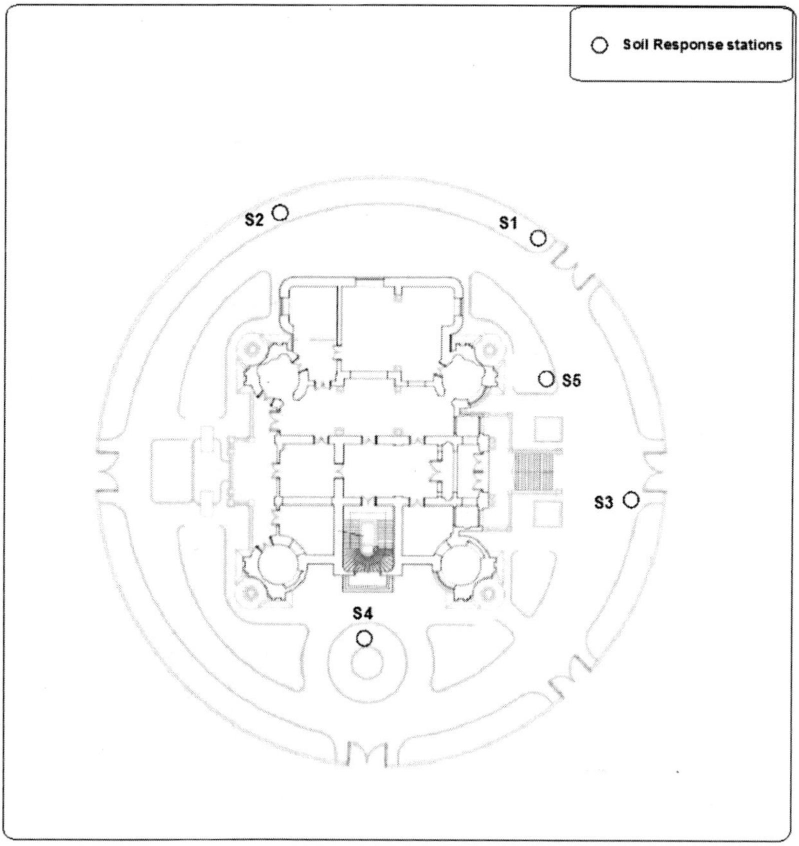

Figure 14. Ambient noise measurement locations

Table 3. Fundamental frequencies and amplification factors at five locations

Site	Fundamental frequency (Hz)	Amplification Factor
S1	3	2.5
S2	3.2	2
S3	3	1.6
S4	3	1.6
S5	3	2

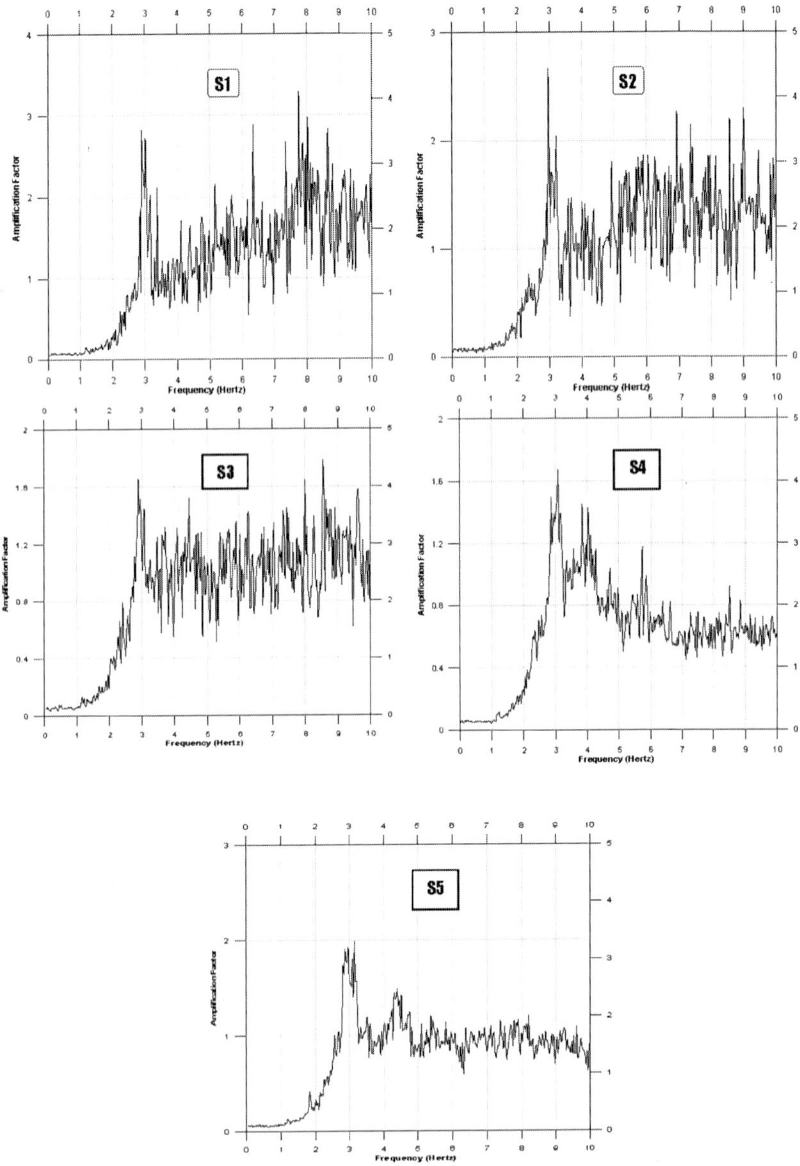

Figure 15. Microtremors soil response for El Sakakini Palace Sites S1 to S5.

Table 4. Parameters used for the Kennett and Kerry method (1979)

Thickness	P-wave velocity (m/s)	S-wave velocity (m/s)	Dry Density (gm/c.c)	Quality factor Qs
10	1300	315 (350?)	1.6	7
25	1300	500	1.7	15
>10.5	2000	700	2	100
	3000	1200	2.5	200

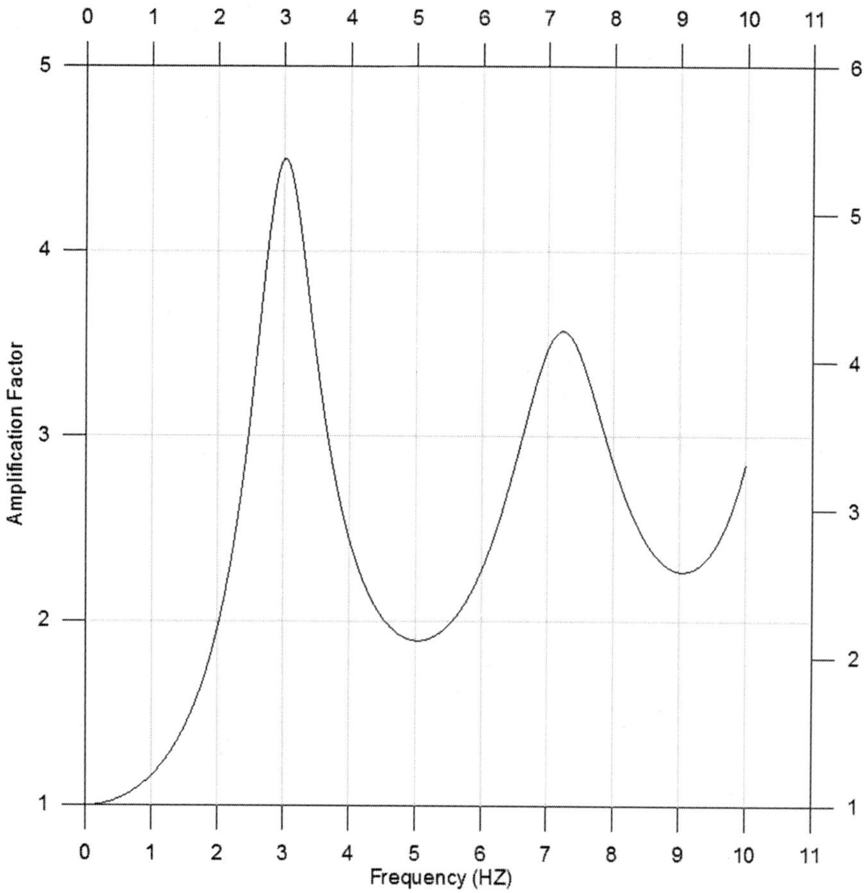

Figure 16. Theoretical ground response analysis at EL Sakakini Palace using Kennett at al. (1979) method.

5.3. Building response

The El Sakakini building is composed of a basement and five floors the upper two being wooden. Figures 17 to 19 show the locations of recording stations used to drive El Sakakini building response. Figures 21 to 26 and Table 5 show the recorded natural frequency of vibration for each floor. All floors show nearly the same resonance frequency with the soil (3-4 HZ). The wooden floors (Figure 25 & 26) show very high amplification and multi peak as fundamental and other harmonics. The fundamental natural frequency of vibration is always the most important frequency that insert the maximum earthquake vibration energy into structure. However when we find other mode of vibrations with big amplification factors we consider this as a warning that this structure may suffer from vibration. This could be very good warning for its unstable performance during vibration.

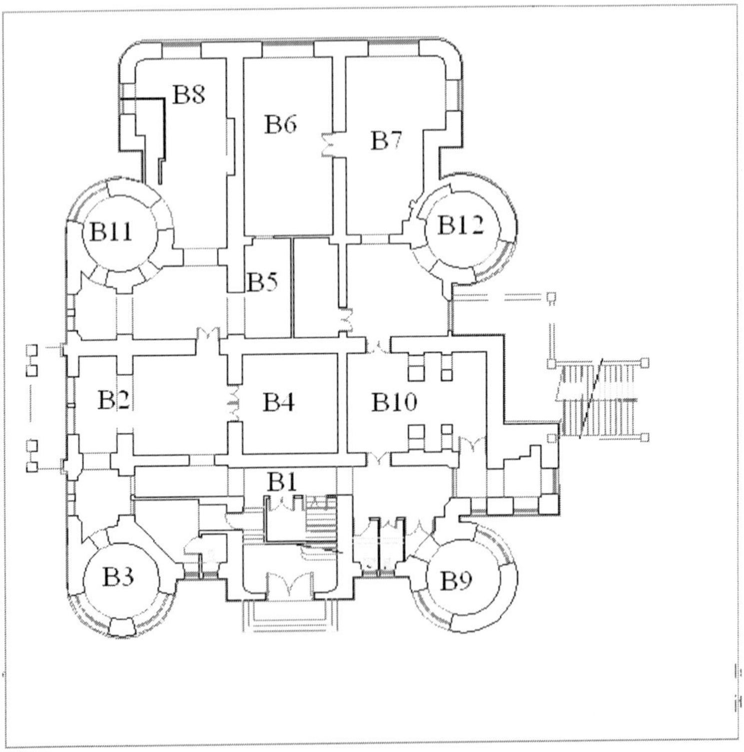

Figure 17. Location of stations at the basement of EL-Sakakini Palace.

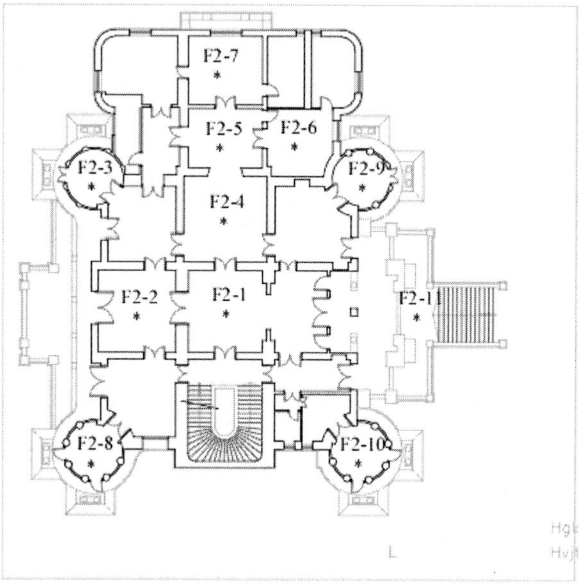

Figure 18. Location of stations at 2nd floor of El-Sakakini Palace.

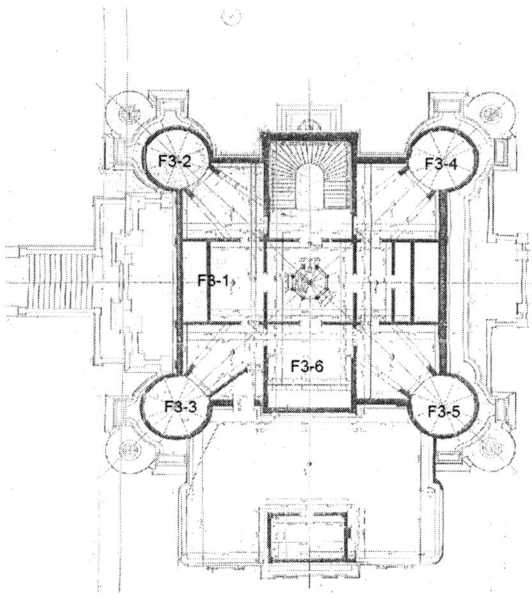

Figure 19. Location of stations at the 3rd floor.

Figure 20. High dynamic range ES-3000 Geometrics mobile station and triaxial geophone used 4 Hz to drive structure response of El-Sakakini Palace.

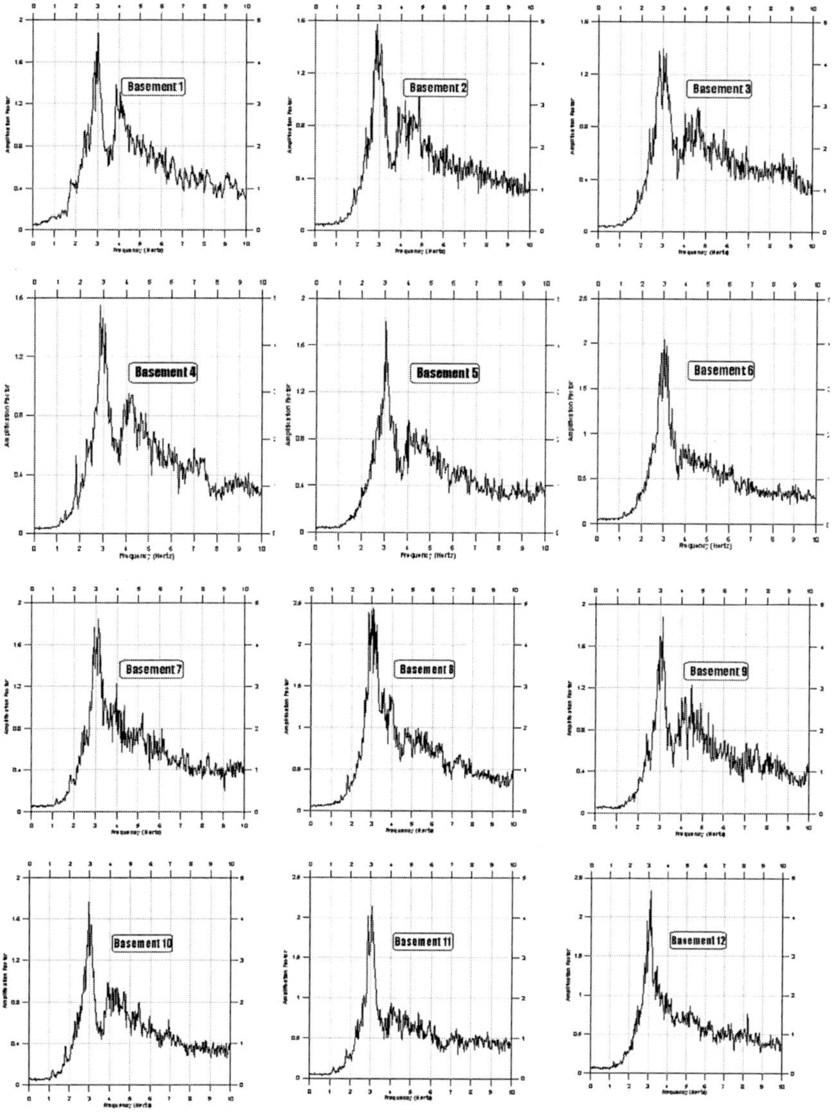

Figure 21. Natural frequency of vibration for basement floor.

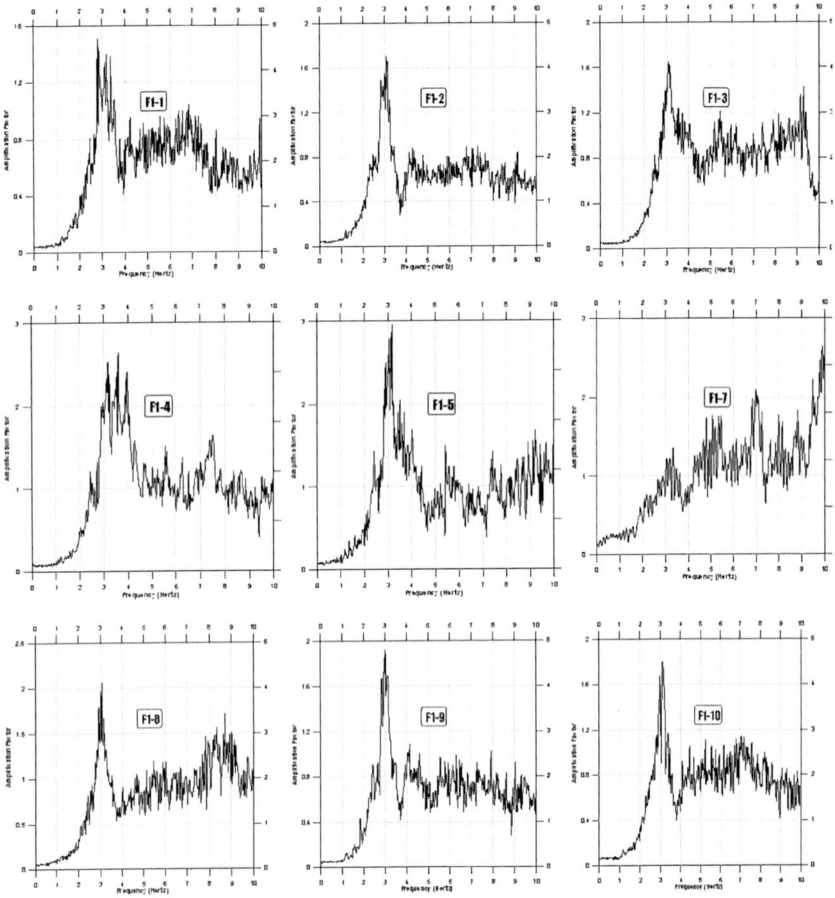

Figure 22. Natural frequency of vibration for the 1st floor.

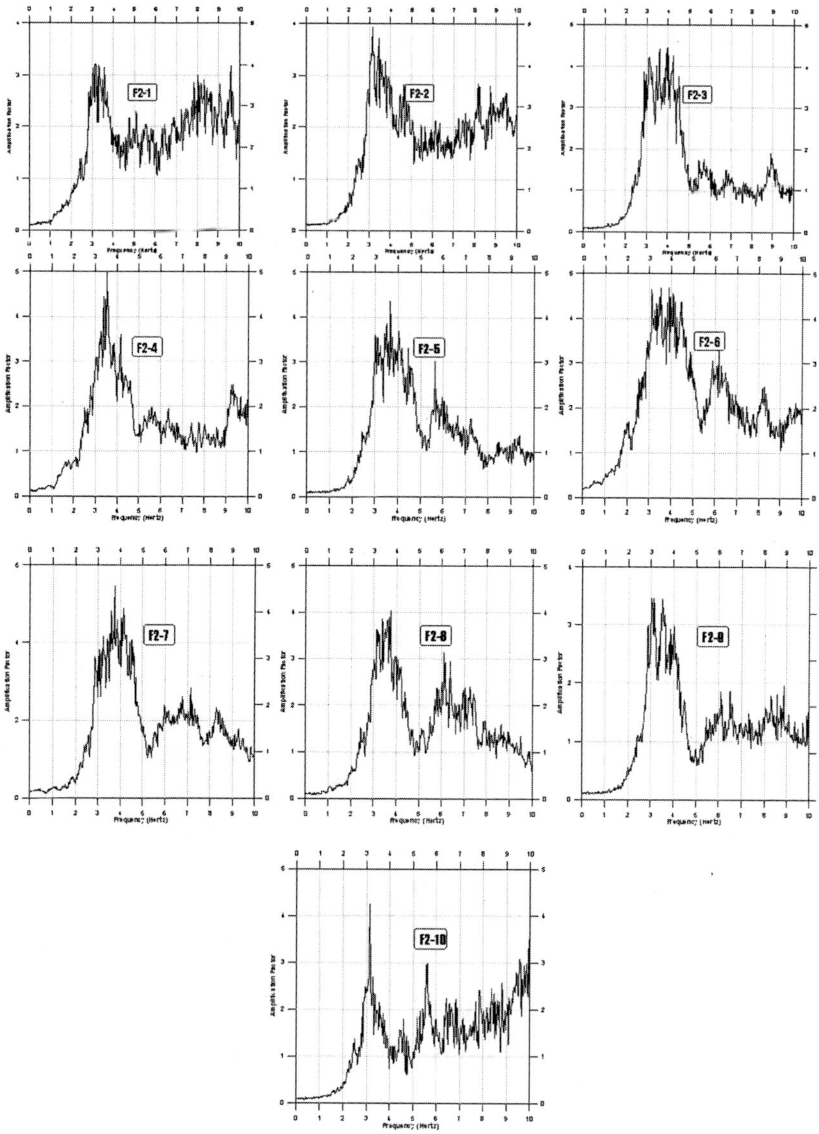

Figure 23. Natural frequency of vibration for the 2nd floor.

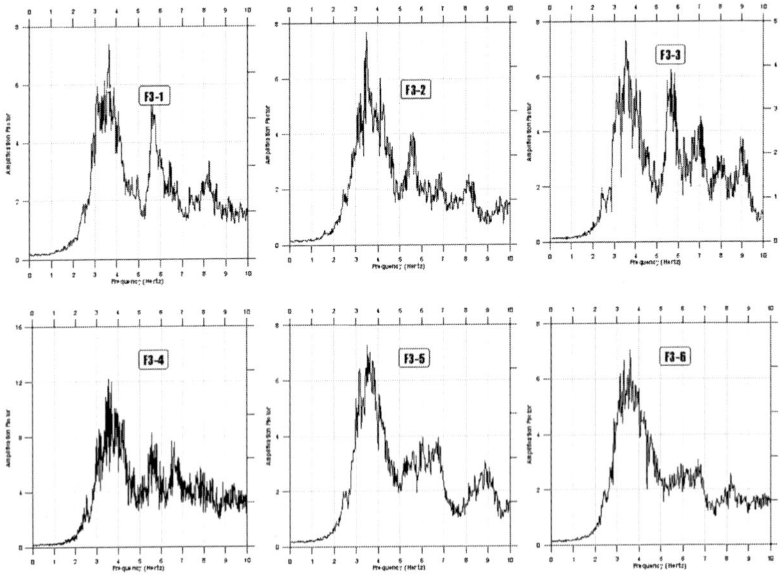

Figure 24. Natural frequency of vibration for the 3rd floor.

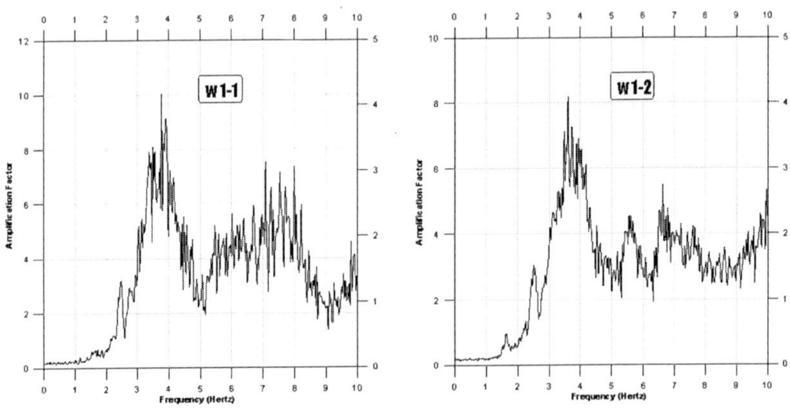

Figure 25. Natural frequency of vibration of the 4thfloor (wooden).

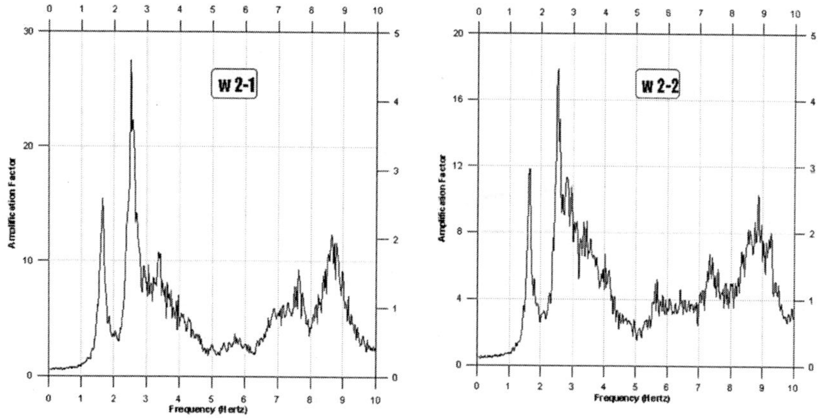

Figure 26. Natural frequency of vibration of the 5^{th}floor (wooden).

Table 5. Natural frequencies of vibration of El Sakakini Palace.

Floor	Fundamental resonance frequency (HZ)	Amplification Factor
Basement	3	1.3-2
1st Floor	3-3.5	1.4-2.8
2nd Floor	3-4	3-4.5
3rd Floor	3-4.3	5.5-8
4th wood Floor	3-4.5	7.5-8
5th wood Floor	2.5	12-20

6. CONCLUSIONS

ElSakakini Palace is an important monument in Egypt. We presented the main results of the seismic hazard analysis and the geophysical campaign to estimate the main characteristics of the ground response and the structure. Based on the available maximum intensity maps for historical earthquakes (>2200BC) the maximum Mercalli Intensity expected at ElSakakini Palace site is VII.

The peak horizontal acceleration at the seismic rock basement found at -35m approximately, and for 10% probability of exceedance in 50 years is 144 cm/sec^2 (0.147g), while for 100 years is 186 (cm/sec^2) (0.19 g). We determined the average soil profile using different geophysical campaigns. It is found that the upper layer has an average shear wave velocity lower than 300m/s and a thickness of 5 to 10meters. It is a man made fill material in rather loose conditions. Below there is a clayey material with average Vs velocity equal to 400-600m/s. At -35m in average we found saturated compacted sand and gravels with Vs velocity exceeding 700m/s. It is considered as the seismic bedrock for the foreseen detailed site-specific analysis of the ground response.

Based on the ambient noise campaign the fundamental frequency of the ground is of the order of 3.0 to 3.5sec very close to the fundamental frequency of the palace. Resonance phenomena should be expected and considered seriously in the detailed analysis of the structure. There are strong evidences that the upper two stories with wooden floors, which are presenting high amplification factors, are subjected to several damages and degradation of their bearing capacity.

REFERENCES

1. Apostolidis P., Raptakis D., Roumelioti Z., Pitilakis K., (2004), Determination of S-Wave velocity structure using microtremors and SPAC method applied inThessaloniki (Greece). Soil Dynamics & Earthquake Engineering, 24, 1, 49-67.
2. Ben-Menahem, A., (1979). Earthquake catalogue for the Middle East (92 B. C. - 1980 A. D.): Bollettinodi Geofisica Teoricaed Applicata, 21, No. 84, 245 - 310
3. Castro, R. R., F. Pacor, A. Sala, and C. Petrungaro (1996). S Wave attenuation and site effects in the region of Friull, Italy, J. Geophysics. Res. 101, 22355-22369.
4. Celebi, M., C. Dietel, J. Prince, M. Onate, and G. Chavez (1987). Site amplification in Mexico City (determined from 19 September 1985 strong-motion records and from records of weak motion), in Ground Motion and Engineering Seismology, A. S. Cakmak (Editor), Alsevier, Amsterdam, 141-152.
5. Chopra, A. K. (1981). Dynamics of structures, a primer Earthquake engineering research institute.
6. Chopra, A. K., (1995). "Dynamic of structures – Theory and Application to Earthquake Engineering, Englewood Cliffs, New Jersey.

7. Clayton, R. W., and McMechan, G. A., 1981, Inversion of refraction data by wavefield continuation: Geophysics, v. 46, p. 860-868.

8. Egyptian Code. Egyptian code of Practice (ECP-1993) for estimating loads and forces, 1993.

9. F. Nateghi-A, (1994). Seismic strengthening of a ten story steel framed hospital: Earthquake resistant Construction & design, Savidis (ed.), Balkema, Rotterdam, ISBN 90 5410 392 2, P 849-856.

10. Field E. H., S. E. Hough, and. H. Jacob (1990). Using microtremors to assess potential earthquake site response, a case study in Flushing Meadows, New york city. Bulletin of the Society of America, Vol. 80. No. 6, pp. 1456-1480, December, 1990.

11. Gamal M. A. and Noufal A., A Neotectonics model AND seismic hazard assessment for Egypt, 8 th international conference on geology of the Arab world, Cairo university Egypt 13-16 2006.

12. Jordan Sesimoloical observatory (1998-2000), national resources authority geophysics and technical services, bulletin (1998-2000).

13. Joyner, W. B., and D. M. Boore, 1981, Peak horizontal acceleration and velocity from strong motion record including records from the 1979 Imperial Valley, California, earthquake, Bull. Seism. Soc. Am., 17, 2011-2038.

14. Kagami, H., C. M. Duke, G. C. Liang, and Y. Ohta (1982). Observation of 1-5 second microtremors and their application to earthquake engineering. Part II. Evaluation of site effect uppon seismic wave amplification due to extremely deep soils., Bull. Seism. Soc Am. 72, 987-998.

15. Kagami, H., S. Okada, K. Shiono, M. Oner, M. Dravinski, and A.K. Mal (1986). Observation of 1-5 seconds microtremors and their application to earthquake engineering. Part III. A two-dimensional of the site effect in San Fernando valley, Bull. Seism. Soc. Am. 76, 1801-1812.

16. Kanai, K. (1957). The requisite conditions for predominant vibration of ground, Bull. Earthq. Res. Inst., Tokyo University, 31, 457.

17. Kanai, K. (1962). On the spectrum of strong earthquake motions, Primeras J. ArgentinasIng.Antisismica 24, 1.

18. Kanai, K. and Tanaka (1961). On microtremors. VII, Bull. Earthq. Res. Inst., Tokyo Univ., 39, 97-115.

19. Langston, C. A. (1977). Corvallis, Oregon, Crustal and upper mantle receiver structure from teleseismic P and S waves, Bull. Seism. Soc. Am. 67, 713-724.

20. Lermo, J. and F. J. Chavez-Garcia, (1993). Site effect evaluation using spectral ratios with only one station, Bull. Seism. Soc. Am. 83, 1574-1594.

21. Lermo, J., M. Rodriguez, and S. K. Singh (1988). The Mexico Earthquake of September 19, 1985: Natural periods of sites in the valley of mexico from microtremorsmeasurments and from strong motion data, Earthquake Spectra 4, 805-814.

22. Louie, J, N., 2001, Faster, Better: Shear-wave velocity to 100 meters depth from refraction microtremors arrays: Bulletin of the Seismological Society of America, v. 91, p. 347-364.

23. Maamoun, M., 1979:Macroseismic observations of principal earthquake in Egypt, HelwanInstitue of Astronomy and Geophysics, Bulletin 183, 120.

24. Maamoun, M., A. Megehed, and A. Allam, (1984).Seismicity of Egypt, Institute of Astronomy and Geophysics 4.

25. Makropoulos, K. C., and P. W. Burton, (1981).A catalogue of Seismicity in Greece and adjacent areas.Geophys. J. R. Astr. Soc. 65, 741 - 762.

26. Manakou M., Raptakis D., Apostolidis P., Chávez-García F. J., Pitilakis K., (2010), 3d soil structure of the Mygdonian basin for site response analysis, Soil Dynamics and Earthquake Engineering, vol. 30, pp. 1198-1211

27. McGuire, R.K., (1993), FORTRAN computer program for seismic risk analysis. U.S. Geol. Surv.

28. Miller, R. D., Park, C. B., Ivanov, J. M., Xia, J., Laflen, D. R., and Gratton, C., 2000, MASW to investigate anomalous near-surface materials at the Indian Refinery in Lawrenceville, Illinois: Kansas Geol. Surv. Open-File Rept. 2000-4, Lawrence, Kansas, 48 pp.

29. Mohamed A. Gamal, A. Noufal and E. Ghoneim, ANeotectonics model and seismic hazard assessment for Egypt (2006): (Accepted for publication) 8th international conference of geology of the Arab world, 13-16 Feb. 2006 Cairo university Giza, Egypt.

30. Mucciarelli M., (1998). Reliability and applicability of Nakamura's technique using microtremors: an experimental approach, Journal

of Earthquake Engineering, Vol.2, No.4 625-638 Imperial college press.

31. Nakamura, Y. (1989). A method for dynamic characteristic estimation of subsurface using microtremors on the ground surface. Q. R. Railway tech. Res. Inst. Rept. 30, 25-33.

32. NEIC, (1973-2006). Prelimenary determination of Epicenters, monthly listing, U.S. Department of the interior/ Geological survey, National Earthquake information center:.

33. Nogoshi, M. and T. Igarashi (1970).On the propagation characteristics of microtremors, J. Seism. Soc. Jpn. 23, 264-280.

34. Pitilakis K., Anastasiadis A., Kakderi K., Manakou M., Manou D., Alexoudi M., Fotopoulou S., Argyroudis S., Senetakis K., 2011, "Development of comprehensive earthquake loss scenarios for a Greek and a Turkish city: Seismic hazard, Geotechnical and Lifeline Aspects", Earthquakes and Structures,Vol. 2, No 3, September 2011.

35. Pullammanappallil, S., Honjas, B., Louie, J., Siemens, J. A., and Miura, H., 2003, Comparative study of the refraction microtremors method: Using seismic noise and standard P-wave refraction equipment for deriving 1D shear-wave profiles, Proceedings of the 6th SEGJ International Symposium (January 2003, Tokyo), 192-197.

36. Raptakis D., Manakou M., Chavez-Garcia F., Makra K., Pitilakis K., (2005), 3D configuration of Mygdonian basin and preliminary estimate of its site response,Soil Dynamics and Earthquake Engineering, 25, 871-887.

37. Riad, S., and H. Meyers, (1985).Earthquake catalog for the Middle East countries 1900 - 1983. World Data Center.

38. Rogers, A. M., R. D. Borcherdt, P. A. Covington, and D. M. Perkins (1984). A comparative ground response study near Los Angeles using recordings of Nevada nuclear tests and the 1971 San Fernando earthquake, Bull. Seism. Soc. Am. 74, 1925-1949.

39. Shapira, A., (1994).Seimological Bulletin of Israel. 1900 - 1994.

40. Thorson, J. R., and Claerbout, J. F., 1985, Velocity-stack and slant-stack stochastic inversion: Geophysics, v. 50, p. 2727-2741.

41. Uniform Building Code (UBC,1985). International Conference of building officials whittier, California.

42. Wilson, E.L, and Habiballah, A (1990)."SAP2000, Structural Analysis Programe" computer and structures Inc., Berkeley, California.

43. Woodword Clyde Consultant WWCC, (1985).Earthquake activity and dam stability for the Aswan High Dam, Egypt.High and Aswan Dames Ministry of Trregation, Cairo 2.

2

CHAPTER

SEISMIC AMBIENT NOISE AND ITS APPLICABILITY TO MONITOR CRYOSPHERIC ENVIRONMENT

Won Sang Lee[1], Joohan Lee[1] and Sinae Han[1, 2]

[1] Korea Polar Research Institute, Republic of Korea

[2] Kangwon National University, Republic of Korea

1. INTRODUCTION

In seismology, most of researches have been investigated by analyzing major seismic 'signals', for instance, body waves and surface waves. Aki [1] first introduced 'coda' waves, which had long been recognized as 'noise', consisting of scattered *S*-waves during propagation through the heterogeneous Earth media. Since then, a number of studies have been conducted to measure medium heterogeneity using coda waves over the world (e.g. [2]). Another revolutionary research dealing with 'noise' in seismology has been reported by [3]. The authors introduced a remarkable method to determine surface wave velocity examining long sequences of seismic ambient noise. It gives us a great opportunity to explore velocity structures underneath by nothing but listening to noise.

An additional interesting feature in terms of 'noise' shown up in the broadband seismic records is microseisms having two predominant peaks in a frequency domain such as primary and secondary microseisms, which have been believed to be originated by long-period ocean waves. The most widely accepted mechanisms for the generation of microseisms are as follows: (1) When ocean waves impact the coast, a part of acoustic energy is transferred into the crust. The directly converted seismic energy (Primary Microseisms, PM) from ocean waves propagates mostly as Rayleigh waves having a predominant period near 8-20 s which is the same period as the ocean waves even *P*-waves have been observed [4]. (2) The most energetic ambient noise is referred to as the secondary microseisms, or Double-Frequency (DF) microseisms, with 4-10 s of predominant period and the generation mechanism is more complex than that of PM. As ocean waves travel toward and strike the coast, reflected waves are generated and nonlinearly interact with incident waves in shallow regions, which results in a frequency doubling of a standard ocean wave [5,6]. The pressure amplitude of propagating incident waves decays exponentially as water depth becomes deeper, whereas the amplitude of standing waves keeps nearly constant with depth. Such powerful DF microseisms could be efficiently excited by significant reflection of wave energy at steep coastlines.

There have been several research efforts (e.g. [7,8]) to identify the source regions of DF microseisms utilizing array analysis, which report that the most DF microseisms are excited near the coastal regions where the swell reaches steep coasts with normal incidence, in good agreement with the Longuet-Higgins model for the generation of DF microseisms [7]. In addition to the determination of source location, scientists have investigated if the DF microseisms vary seasonally in amplitude. According to the spectral analysis for seismic ambient noise by McNamara and Buland [9], the power of DF noise levels in winter is higher than the summer season, which could be observable over the northern hemisphere. Since the DF microseism has been originated by ocean waves, it tends to show seasonal variability, reflecting the vigor of ocean activities [10]. This phenomenon allows us to monitor the Earth's near-surface environment in the ocean using the ambient noise analysis. In Polar Regions, seasonal variation in the DF energy occurs inconsistently compared to the characteristics shown in the lower latitude regions. Recent observations (e.g. [11,12]) suggested that there is a possible relation between the seasonal variability of DF microseisms and sea-ice variability. Tsai and McNamara [13] has theoretically explained that the sea-ice concentration is responsible for the seasonal change of DF microseisms in power with a simple attenuation model.

Even though many literatures have reported fascinating results for seismic ambient noise in both the northern and southern hemispheres, there are only a few studies regarding the DF microseisms in Antarctica [11,12,14,15] due to a dearth of broadband seismic stations. In this chapter, we combine more 3-year data from 2009-2011 with the previous results [11] and present a seismic ambient noise study that could provide more reliable evidence to present strong association between the seasonal change of DF and the variability of sea-ice concentration, in turn we are able to monitor regional cryospheric environment.

2. DATA AND SPECTRAL ANALYSIS

Korea Polar Research Institute (KOPRI) has been operating a permanent broadband seismic station (KSJ1, 62.22ºS/58.78ºW; Figure 1) at the King George Island (KGI) in the South Shetland Islands, Antarctica, since 2001. The seismic station is mainly equipped with a three-component broadband Streckeisen seismometer (STS-2) and a 24-bit high-resolution data logger (Q4124). The sensor responses in amplitude and phase guarantee that signals recorded within the frequency range between 120 sec and 20 Hz are reliable to be used without severe distortion. We have collected seismic data with 1 (LHX) and 20 Hz (BHX) sampling rates, and used the BHZ data for the spectral analysis.

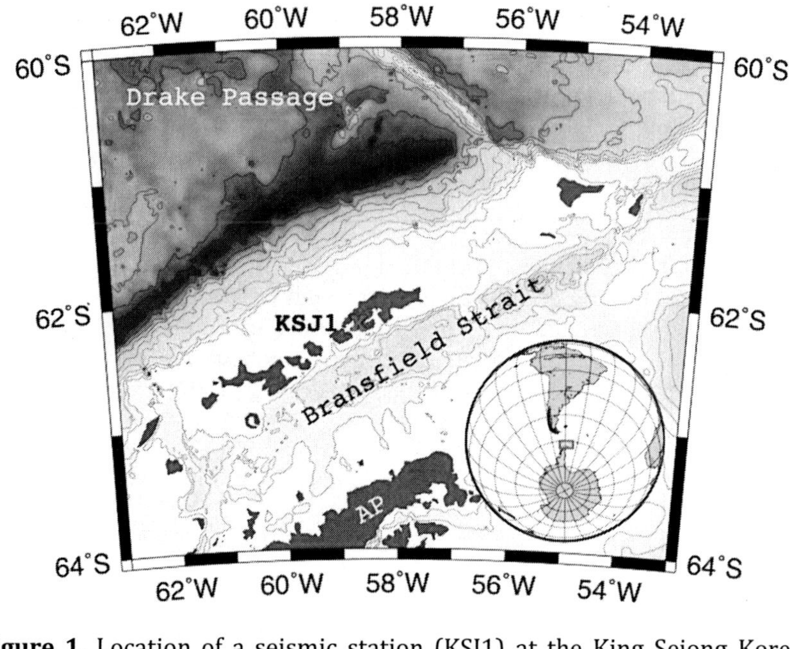

Figure 1. Location of a seismic station (KSJ1) at the King Sejong Korean Antarctic Base in the King George Island (KGI) marked by a red X. The Island is situated in between the Drake Passage and the Bransfield Strait near the Antarctic Peninsula (AP).

In order to examine the spectral characteristics of seismic ambient noise, we calculate the Power Spectral Density (PSD) of the seismic noise following the rigorous method by McNamara and Buland [9]. The method requires first parsing continuous time series into 1-hour time series segments, overlapping by 50 % and distributed continuously throughout the day, week, and month. The PSD estimate should be converted into decibels with respect to acceleration (meters/second^2)^2/Hertz for direct comparison to the standard noise model [16] in this study. The PSD technique provides stable spectra estimates over a broad range of periods (0.05–100 sec); however, it suffers from poor time resolution due to the long transforms (3600 sec) and requires many hours of data to compile reliable statistics. For better resolution at shorter periods, a larger number of shorter records should be analyzed [9]. From more than 90,000 PSDs for the period of 2006-2011, we could estimate Probability Density Functions (PDFs) to investigate the highest probability noise level (mode) for each channel as a function of period. As the method utilizes modes rather than higher energy level, we could obtain more reliable information to understand the characteristics of ambient noise, since even damaging earthquakes occurred near the station it is just a small portion out of background noise in terms of occurrence [9]. Moreover, it has a distinctive feature that we do not need to screen continuous quiet time window. At present, it has been known as the most common and robust technique to measure seismic ambient noise and evaluate the performance of seismic stations. More details regarding statistics and spectral analysis should be referred to [9].

Figure 2 demonstrates a statistical view of broadband PDFs of a vertical component for the period of 2006-2011. Two prominent peaks show up around 5 and 10 s in period, which correspond to secondary and primary microseisms, respectively. HNM and LNM indicate (gray curves in Fig. 2) the standard high and low noise model [16], respectively. Although several earthquakes occurred near the station during the operation period, they do not affect the overall PDFs as we mentioned earlier.

3. SEASONAL VARIABILITY OF DF MICROSEISM AND ITS ASSOCIATION WITH SEA-ICE CONCENTRATION

We investigated spectral characteristics for KSJ1 through the estimation of PSDs and PDFs of seismic background noise (Fig. 2), and found that the primary and secondary microseisms appear distinctly on the plot.

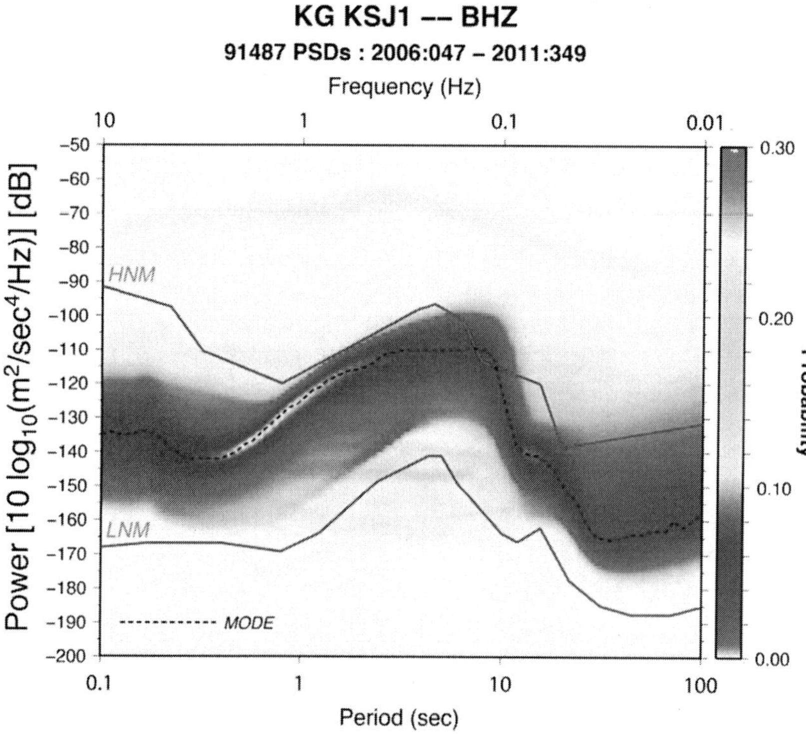

Figure 2. A Probability Density Function (PDF) plot of BHZ for KSJ1 during 2006-2011. Two predominant peaks appear around 5 and 10 s in period, corresponding to secondary (or DF) and primary microseisms, respectively. HNM and LNM indicate (gray curves) the standard high and low noise model (Peterson, 1993), respectively. The most probable energy with respect to frequency is presented by a dashed curve (mode).

Comparing to the HNM and LNM, we may evaluate that KSJ1 has been well operated in terms of system performance except slightly noisier (or might be higher energy level of DF microseisms) than the HNM around 10 s. In general, a plot of PDFs could provide helpful information on spectral signature of a station; however, temporal patterns of the microseisms are barely identifiable from it. In an attempt to examine temporal variation of the noise level of KSJ1, we obtain the statistical mode for the corresponding periods from daily PSDs so that we could construct a power spectrum with respect to time (Fig. 3). Empty spaces in the figure show data missing due to most likely system malfunctioning. There is nothing noticeable in the period of longer than 10 s throughout the operation time. Having interests in the feature near 4-10 s in period, i.e. DF microseisms, it happens that the DF energy comes to be weaker from July through September (austral winter). The behavior is apparent in 2007, 2009, and 2011, whereas it becomes ephemeral in 2006, 2008, and 2010, but rather weaker in power compared to other seasons in a year. This observation contrasts with the seasonal variability of seismic noises in the northern hemisphere; for instance, the amplitude of the Earth's hum reaches its seasonal maximum in winter season [17,18] revealed from an array analysis. The power of DF microseisms in the northern hemisphere shows a similar pattern (e.g. [19]) as the Earth's hum. Most literatures suggest that these characteristics are attributed to seasonal variation of the intensity of infragravity wave depending on swell amplitudes.

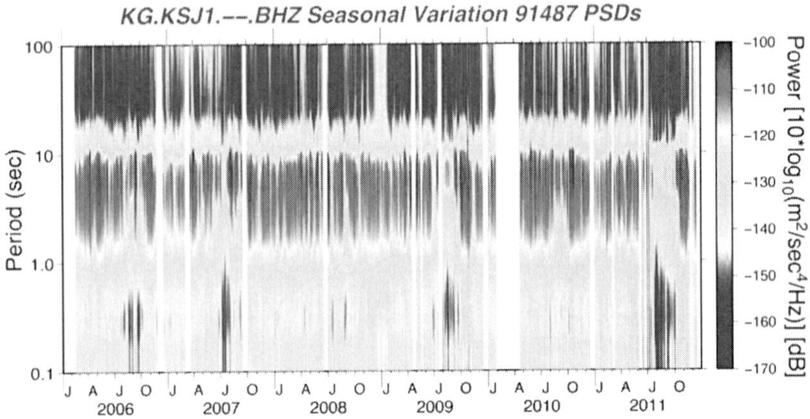

Figure 3. Spectral amplitude variation in seismic noise for BHZ (broadband vertical component) during the period of 2006-2011. Note that the seismic energy at the frequency range of DF microseisms (4-8 s) becomes weaker during July to September annually, which is a different behavior from that of the northern hemisphere except the Arctic region. Empty spaces in the plot indicate data missing or a period of malfunctioning.

Ringdal and Bungum [20] reported a pure sinusoidal pattern in long period noise level, i.e. seasonal maximum in winter and minimum in summer, from a spectral analysis of NORSAR data for three years. It does not, however, necessarily occur in the Polar Regions, especially Antarctica, and might be due to a regional difference between the northern and southern hemispheres. More specifically, [15] similarly observed weaker energy of DF microseisms in austral winter at the station DRV, Antarctica, and explained that the acoustic energy from ocean swell tends to be severely attenuated by sea ice and reflecting waves along the coast suffers as well causing fewer DF microseisms generated by sources. We refer to it as 'sea-ice damping effect' in this study. Recently, numerical modeling approach to figure out this phenomena has been made by Tsai and McNamara [13], which shows that 75-90 % of the variability in microseism power in the Bering Sea can be predicted using a simple model of microseism damping by sea ice. Moreover, they argued that we could use the microseism as a good indicator to monitor the strength of sea ice that is not easily measured by through other means.

In order to carefully study the direct relation between the energy of DF microseisms and the sea ice condition, we extract and integrate the DF power ranging 4-10 s in period out of the power spectrum and collect Sea-Ice Concentration (SIC) information. To create time series of SIC, we used data based on brightness temperature observations at 89 GHz obtained from the AMSR-E (Advanced Microwave Scanning Radiometer for EOS Aqua) on board NASA's Aqua satellite. The brightness of each image pixel is converted to the SIC using the ASI (ARTIST Sea Ice) algorithm [21]. The data offer SICs on a grid with 6.25 km resolution with a complete daily coverage of the Polar Regions. A domain for calculating SIC covers a part of the Drake Passage allowing us to compare to the DF energy in this study. The calculated percentage of SIC is the percentage of grid cells containing more than 15 % sea ice. This is mainly attributed to the fact that the accuracy of the SIC is ±15 % in regions of first-year ice [22].

As shown in Figure 4, there is clear seasonal variation found in both the power of DF microseisms (red curve) and the SIC (black curve) from 2006-2011. To quantify how they are closely related, we apply cross correlation that is a standard method of estimating the degree to which two series are correlated. The bin size of each time series is chosen to be 1-day. The resultant cross-correlation coefficient is given by -0.70 that is a strong negative correlation. The result implies that as the SIC becomes higher, i.e. more sea-ice in the ocean, the DF power decreases, which is coincident with the hypothesis of 'sea-ice damping effect'. We also determined the lag time as almost zero from the cross correlation, which indicates that the DF energy responses immediately to the sea-ice condition nearby. When one may take a closer look at the period of May through September in a year, it becomes more prominent.

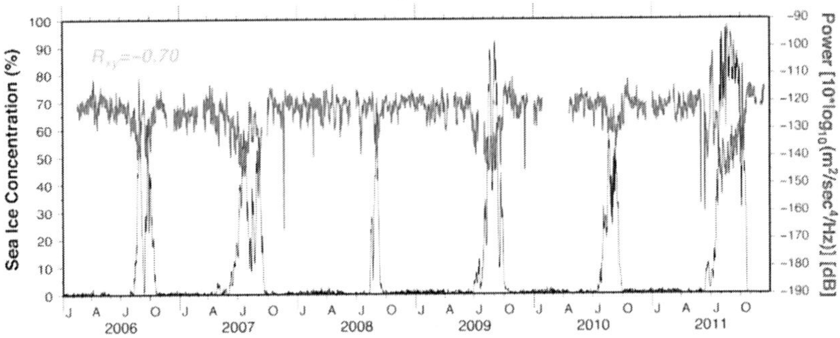

Figure 4. A comparative plot of Sea Ice Concentration (SIC, black curve) sampled near the KGI toward the Drake Passage vs. seismic energy of DF microseisms (red curve) observed at KSJ1 during 2006-2011. When either the SIC increases or decreases, the DF power responses immediately (negatively correlated), suggesting the DF energy is a relevant seismic proxy to monitor cryospheric environment especially the sea ice condition nearby. The strong correlation (-0.70) between them supports the hypothesis.

Remote sensing using satellites allows us to extensively improve our knowledge over various scientific issues, especially in Polar Regions. For instance, we can measure surface melt extent on ice, moving speed of glaciers, ice mass balance by means of AMSR-E, InSAR satellites, and GRACE, respectively. Even though all these methods give us great opportunities to monitor dramatic changes in the cryospheric environment, we should still conduct in-situ measurements to obtain physical and mechanical properties of ice such as stiffness. From the recent development on theoretical approach to predict the variability in DF microseisms [13] and the obvious evidence linking the variation of the DF energy to the SIC in this study, we anticipate that a long-term observation of the DF microseisms could be a good tool to monitor local climate change in Polar Regions. Most of literatures in seismology have dealt with several major issues such as determination of velocity and attenuation structures in the Earth, precise locating techniques, and investigation of earthquake sources. In this study, beyond that horizon, we find out that a seismological method could play a key role in understanding physical interaction between climate change and the cryosphere.

4. CONCLUSIONS

Substantial advances in seismograph technology allow us to consistently observe the Earth's continuous oscillation everywhere in the world. The DF microseism has been known that it is excited by ocean waves, thus it is likely to show seasonal variations [11]. Examining the ambient seismic noise level at KSJ1 during the period of from 2006-2011, we found a distinct seasonal pattern in the period of 4-10 s; the DF energy comes to be weaker from July through September (austral winter) in every year. Cross correlation results tell us that as the SIC becomes higher, the DF power decreases, and confirm that the DF energy responses immediately to the sea-ice condition. Consequently, we propose that a long-term observation of the DF microseisms should be necessary to monitor local climate change in Polar Regions, which contributes extra benefits to the satellite remote sensing.

ACKNOWLEDGEMENTS

This research has been supported by KOPRI research grants PN12040 (CATER 2012-8080) and PE13050.

REFERENCES

1. Aki K. Analysis of seismic coda of local earthquakes as scattering waves. Journal of Geophysical Research 1969; 74: 615-631.
2. Sato H, Fehler MC. Seismic Wave Propagation and Scattering in the Heterogeneous Earth. AIP Press/Springer Verlag; 1998.
3. Shapiro NM, Campillo M, Stehly L, Ritzwoller MH. High-Resolution surface wave tomography from ambient seismic noise. Science 2005; 307: 1615- 1618.
4. Gerstoft P, Shearer PM, Harmon N, Zhang J. Global P, PP, and PKP wave microseisms observed from distant storms. Geophysical Research Letters 2008; 35 L23306:
5. Longuet-Higgins MS. A theory of the origin of microseisms. Philosophical Transactions of the Royal Society A 1950; 243: 1 –35.

6. Hasselmann K. A statistical analysis of the generation of microseisms. Reviews of Geophysics 1963; 1: 177– 210.

7. Chevrot S, Sylvander M, Benahmed S, Ponsolles C, Lefèvre JM, Paradis D. Source locations of secondary microseisms in western Europe: Evidence for both coastal and pelagic sources. Journal of Geophysical Research 2007; B11301:

8. Traer J, Gerstoft P, Bromirski PD, Hodgkiss WS, Brooks LA. Shallow-water seismo-acoustic noise generated by tropical storms Ernesto and Florence. Journal of the Acoustical Society of America Express Letters 2009;

9. McNamara DE, Buland RP. Ambient noise levels in the continental United States. Bulletin of the Seismological Society of America 2004; 94: 1517– 1527.

10. Tanimoto T. Excitation of microseisms. Geophysical Research Letters 2005; 34 L05308:

11. Lee WS, Sheen D-H, Yun S, Seo K-W. The origin of double-frequency microseism and its seasonal variability at King Sejong Station, Antarctica. Bulletin of the Seismological Society of America 2011; 101: 1446-1451.

12. Grob M, Maggi A, Stutzmann E. Observations of the seasonality of the Antarctic microseismic signal, and its association to sea ice variability. Geophysical Research Letters 2011; 38 L11302.

13. Tsai VC, McNamara DE. Quantifying the influence of sea ice on ocean microseism using observations from the Bering Sea, Alaska, Geophysical Research Letters 2011; 38 L22502.

14. Hatherton T. Microseisms at Scott Base. Geophysical Journal of the Royal Astronomical Society 1960; 3: 381-405.

15. Stutzmann E, Roult G, Astiz L. GEOSCOPE station noise levels. Bulletin of the Seismological Society of America 2000; 90: 690–701.

16. Peterson J. Observation and modeling of seismic background noise. U.S. Geological Survey Technical Report 1993; 93(322): 1-95.

17. Rhie J, Romanowicz B. Excitation of Earth's continuous free oscillations by atmosphere-ocean-seafloor coupling. Nature 2004; 431: 552–556.

18. Bromirski PD, Gerstoft P. Dominant source regions of the Earth's "hum" are coastal. Geophysical Research Letters 2009; 36 L13303.

19. Sheen D-H, Shin JS, Kang T-S. Seismic noise level variation in South Korea. Geosciences Journal 2009; 13: 183–190.

20. Ringdal F, Bungum H. Noise level variation at NORSAR and its effect on detectability. Bulletin of the Seismological Society of America 1997; 67: 479-492.

21. Spreen G, Kaleschke L, Heygster G. Sea ice remote sensing using AMSR-E 89-GHz channels. Journal of Geophysical Research 2008; 113 C02S03.

22. Parkinson CL, Coniso JC, Zwally HJ, Cavalieri DJ, Gloersen P, Campbell WJ. Arctic Sea Ice. 1973-1976: Satellite Passive Microwave Observations. National Aeronautics and Space Administration SP-489; 1987.

3

CHAPTER

THE MINARET OF THE GREAT MOSQUE IN ALGIERS, A STRUCTURAL CHALLENGE

Dan Constantinescu, Dietlinde Köber

Krebs & Kiefer Consulting Civil Engineers, Karlsruhe, Germany

ABSTRACT

The Great Mosque in Algiers will be the third largest mosque in the world and its minaret the highest. The region has a high seismic risk. The project designed by a German team of architects and engineers is under construction and will be finished by 2016. Due to the minaret slenderness and to the special composite structure chosen to withstand lateral loading, the structural design faced some challenging aspects. The paper presents the design philosophy, some significant structural features and details of the minaret structure.

Keywords: Earthquake Engineering; High-Rise Buildings; Composite Structures

1. INTRODUCTION

The Algerian state represented by ANARGEMA (Agence Nationale de Réalisation et de Gestion de la Mosquée d'Alger) has commissioned the German Joint Venture KSP/KuK (Consultant Architects and Engineers), winner of an international competition, with the planning of the Great Mosque of Algiers [1]. The building complex is situated in the central axis of the famous Golf of Algiers, facing the Mediterranean Sea, some half way between the old city and the airport and is currently being erected by the China State Construction Engineering Corporation.

The mosque itself covers a surface of 600 m × 150 m [2]. Additional buildings are provided for a cultural center, a library, a religious university and a huge underground parking. The building's complex is seen as the future architectural landmark of the city.

The prayer hall has a squared plane with the side of 150 m, which can accommodate 36,000 prayers and has a central dome with the apex height of 70 m (Figure 1). To mitigate the highly seismic risk of the region, its structure is base isolated by means of a combination of mechanical seismic isolators and hydraulic dampers.

The minaret is a very slender parallelepiped with a total height of 265 m above ground and a squared plane with the side of 26.5 m (Figure 1). Due to this slenderness, to the particularity of the stiffening system and to the strong seismicity of the region, the structural design of the minaret has faced several technical challenges.

The paper presents the design philosophy and some main features of the structure.

2. SEISMICITY

The north of Algeria is a very strong seismic region. Due to the major national importance of the project, the aseis mic design of the minaret has been based on a micro zonation study authored by the Algerian Centre of Applied Research in Earthquake Engineering (CGS).

It recommended the design seismic spectrum given in Figure 2. The spectrum corresponds to a peak ground acceleration of 6.5 m/s^2 and to a return period of 1000 years.

Figure 1. General view of the mosque.

The fundamental period of vibration of the minaret is about 3.7 s and corresponds to a translation. The first torsional period of vibration is 1.1 s.

3. STRUCTURE

The minaret will accommodate a national history and art museum and a corresponding research institute. Over the height of the building there are 5 blocks of 5 stories each separated by sky foyers. The height of each story is 5.85 m and that of a sky foyer is 11.7 m. The transparent top of the minaret envelopes the summit small tower which is typical for Maghreb's region. It is 41 m high and its structure is of steel and glass. There are two underground levels, with the total height of 11.2 m and a squared plane with the side of 50 m. This enlargement of the minaret foot was crucial for ensuring the foundation system.

From the ground level up to the bottom of the summit tube there are four reinforced concrete (RC) cores situated in the corners (Figure 3). They have a squared perimeter with a side varying from 7.75 m at the ground level up to 7.5 m at their top, with variable wall thicknesses over the height and with external walls thicker than the internal ones. They carry out the whole building weight of about 700 MN above ground. The cast-in-place RC floors are designed as girder grids. The main floor beams depicted blue in Figure 3 are also part of the horizontal stiffening system as they couple the cores.

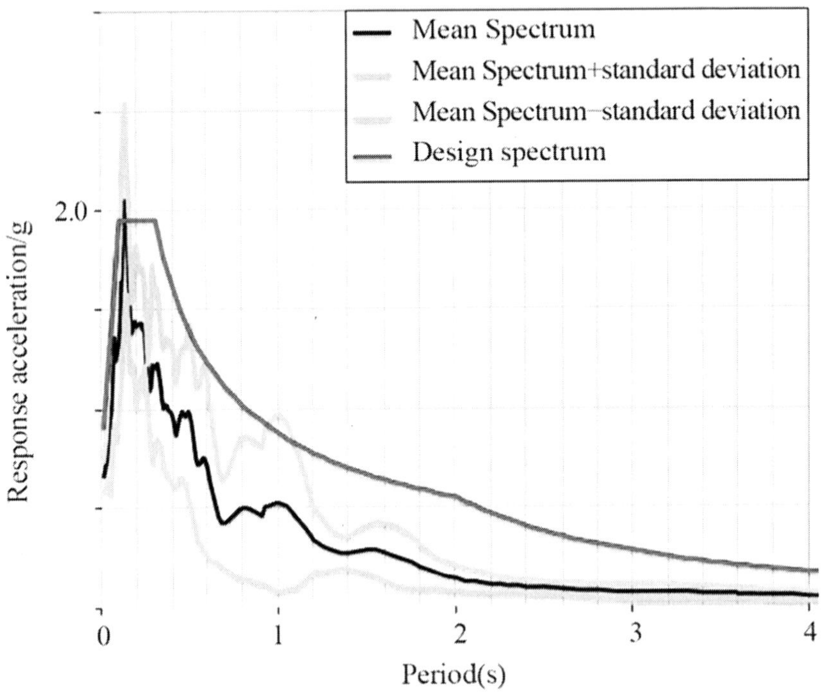

Figure 2. The elastic design spectrum (normalized by means of the gravitational acceleration g = 9.81 m/s²).

The four corner cores have a height to depth ratio of about 30 and are therefore not able to stiffen the tower even if the coupling floor beams are

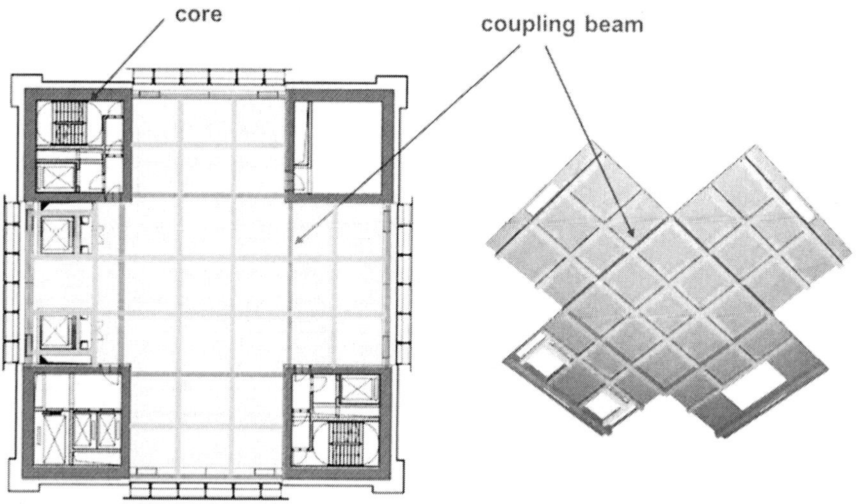

Figure 3. Floor's RC structure

considered. The necessary lateral stiffness and load-bearing capacity can be achieved only if the whole building width is activated. In that case the height to depth ratio becomes about 10. An "outer tube" has to be therefore created. This has been realized by coupling the RC cores by means of X-crossed façade diagonals made of steel sections (Figure 4). On aesthetical grounds the façade diagonals were not desired at the sky foyers, so that a discontinuous bracing has to be used. In order to avoid the transfer of the high internal forces from the coupling steel diagonals to the RC walls and back, a steel construction has been provided within the exterior walls of the cores (see the diagonal and horizontal cross-bars as well as the vertical bars depicted in Figure 4). In this way a composite stiffening system has been created, which combines the RC one made of the corner cores and the coupling floor beams with a spatial steel truss (Figure 5). To accommodate the embedded steel profiles the external walls of the cores have thicknesses varying from 1 m at the ground level and 45 cm towards the top. The corresponding internal walls are 75 and 40 cm, respectively. The coupling effect of the fa- çade bracing is outlined in Figure 6. The overturning bending moment M_0 induced by the seismic action yields internal axial forces N and bending moments M within the cores. Their relative magnitudes depend on the relative stiffness of the two stiffening components, i.e. the cantilevered cores, on one side, and the spatial truss with very stiff flanges, eccentric joints and missing diagonals, on the other side. The truss response to loading is similar to that of a "Vierendeel" beam, except that the coupling between the flanges is realized by means of axial forces within the diagonals instead of bending moments within the connecting members. Due to the relatively high bending stiffness of the cores and to the flexibility of the discontinuous bracing the spatial truss takes over only some 3/4 of the seismic action during an elastic seismic response.

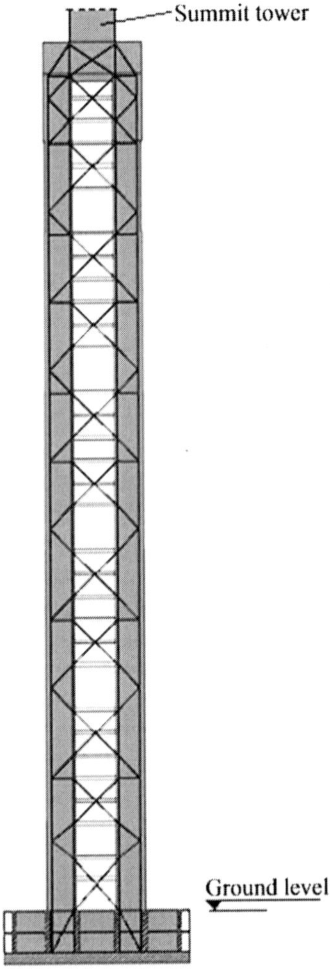

Figure 4. Coupling façade diagonals and steel members cast within the core's external walls.

An additional coupling element has arisen at the top of the tower cores from the walls existing over the height of the last two minaret stories, i.e. 9 m. They are aligned with both the external and the internal walls of the cores. The internal walls have been required to fix the summit tube in the minaret cores. These coupling walls were designed as composite concrete—steel and included steel trusses too, in order to control the concrete cracking during a strong earthquake and ensure a durable ductile be havior. The steel trusses within the external walls are connected with those existing in the core walls (see Figure 4). The steel trusses within the internal walls transfer the internal forces to the RC walls over the height of two stories.

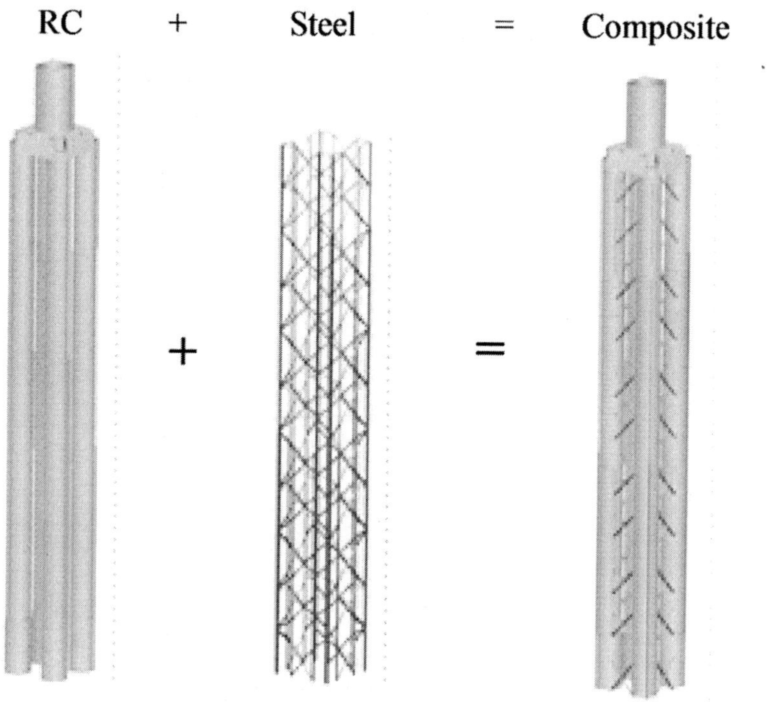

Figure 5. Lateral stiffening system.

The enlargement of the tower foot to the foundation foot required a stiff box over the height of the basement. This has been achieved by means of a grid of RC walls (Figure 7). They were designed to safely carry the vertical and horizontal forces induced at the tower base. The basement walls depicted yellow in Figure 7 are placed beneath the core walls.

The foundation is composed of a 3 m thick foundation slab and 64 “barrettes” (short sheet pile walls). The “barrettes” have a cross-section with the thickness of 1.2 m and the length of 7.2 m at the exterior of the foundation slab and, respectively, 6 m at the interior (Figure 8). Their depth is 43 m.

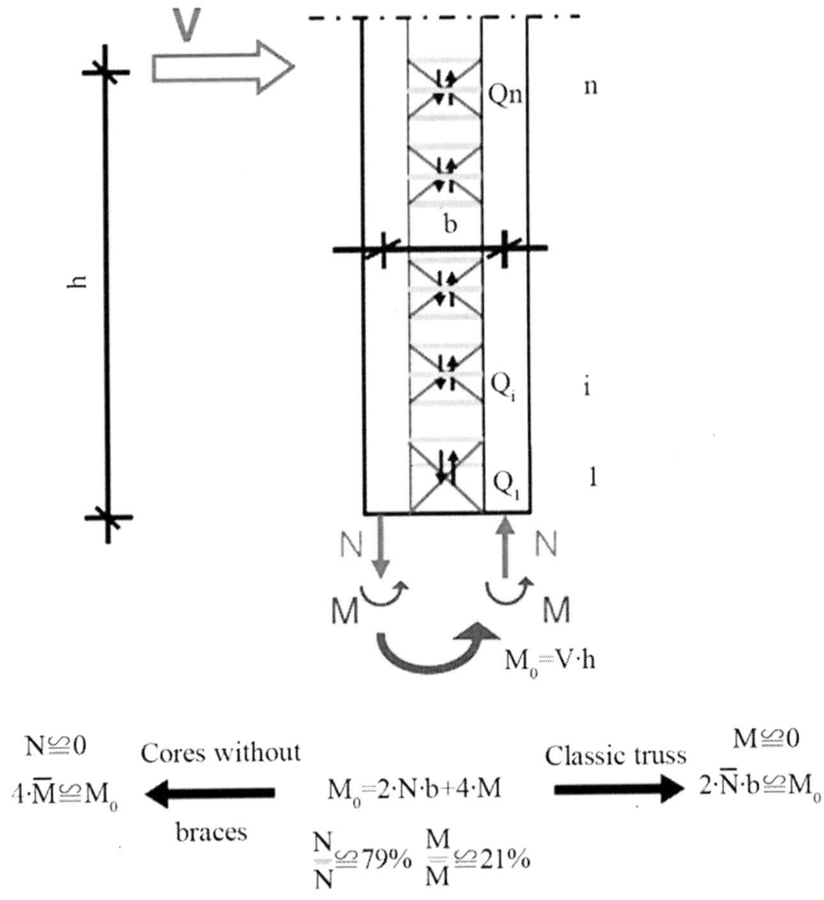

Figure 6. Global effect of the façade bracing.

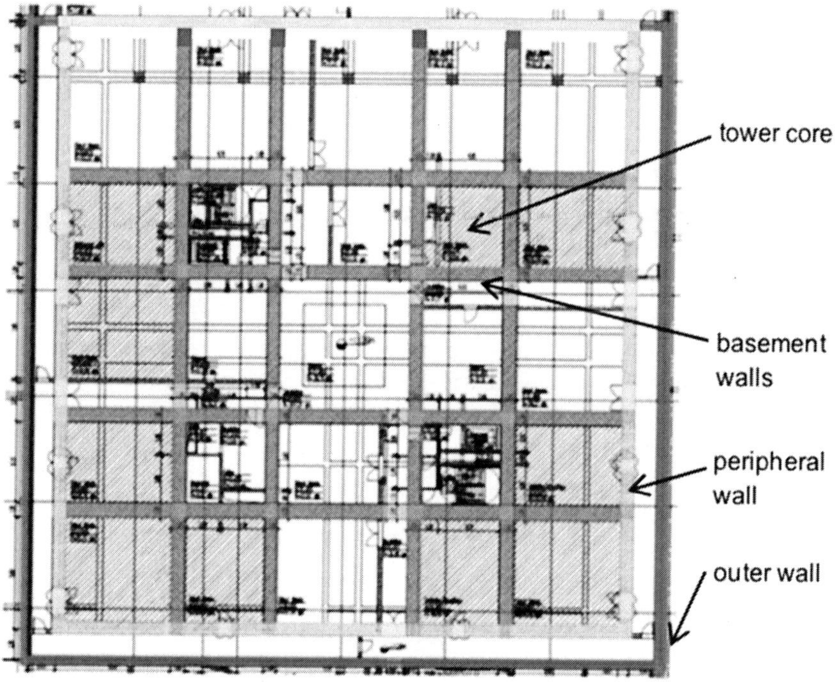

Figure 7. The RC walls within the basement (plan view).

4. DESIGN PHILOSOPHY

The structural design of the minaret tower has been decisively influenced by the existing strong seismic risk and by the client requirement for a millennium lasting monument. The aseismic design of structures has been made in accordance with the performance criteria recommended by [3]. Accordingly, in order to ensure optimal energy dissipation during the design earthquake, one has to protect the "fragile" members by increasing their resistance whereas the "yielding" members have to possess a high ductility.

To ensure a best possible combination of load-bearing capacity and ductility the main structural elements, i.e. the cores, their façade bracing, the basement walls, the foundation slab and the "barrettes", have been designed according to the following categories:

- Highly dissipative members (HDM), i.e. structural elements which will be the first to yield and hence have to possess a high ductility. They will dissipate the most part of the energy induced by a strong earthquake and will also act as "fuses" within the structure by topping the magnitude of the internal forces induced.
- Less dissipative members (LDM), i.e. structural elements which will suffer small, respectively moderate plastic deformation during the design earthquake.
- Elastic members (EM), i.e. structural elements which should remain elastic during the design earthquake. The load bearing capacity of these members should be so scaled, that the higher the risk of a fragile collapse the higher the existing resistance.

The necessary scaling of the members resistances is achieved by applying the method of "design capacity". When two structural members are connected, the less ductile one should have a higher load-bearing capacity as the other.

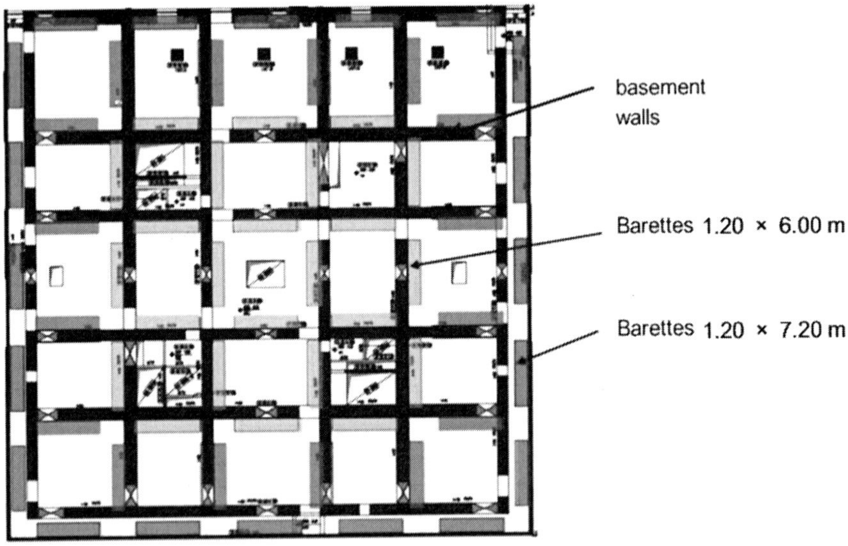

Figure 8. Foundation system (plan view).

This design concept is outlined by the seismic structural response depicted in Figure 9. The magnitude of the internal forces induced by the design earthquake within the structure is dependent on the forces at which the highly ductile members start to yield. Each point on the curve corresponds to a member yielding. The lower the internal force at the first yielding is and the more HDmembers yield, the lower the total force induced by the seism. At the same time the earlier a HD-member enters his plastic range of behavior, the higher should be its ductility. The seismic performance of the structure depends crucially on the safety margin between the maximum horizontal displacement expected to be induced by the design earthquake (in Figure 9 denoted as "displacement demand") and that which the structure is capable to undergo without attending a major disruption.

Figure 10 shows for which category the structural elements of the minaret have been designed. The chosen "fuses" for a strong earthquake are the façade bracings and the main floor coupling beams. In case of the façade bracings only the central parts of diagonals are designed to dissipate energy through plastic deformation. They are made of steel grade S235 which is very ductile and has less strength than the grade S355 used for the other steel profiles. In order to top the induced forces it has been required that the yield strength of S235 must not exceed 245 MPa. The dissipative parts of the façade bracing are bolted with the rest, so that they can be easily replaced after a very strong earthquake, if necessary.

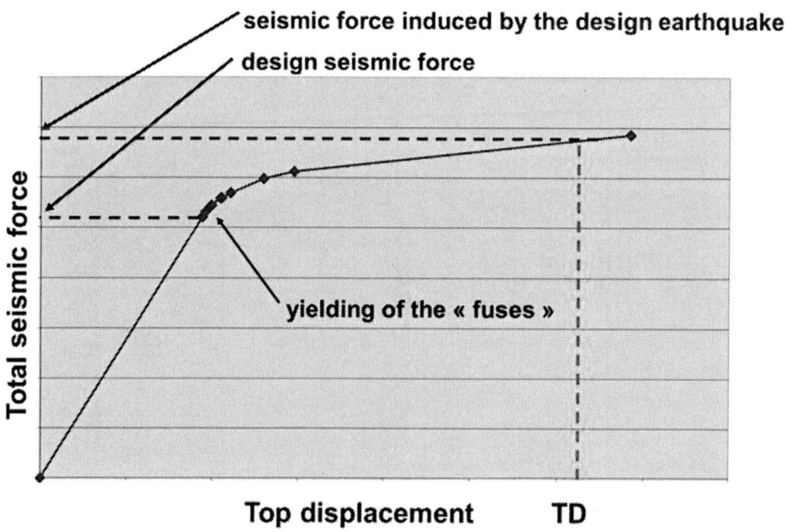

Figure 9. Relationship between the base seismic force and the top displacement (TD = displacement demand).

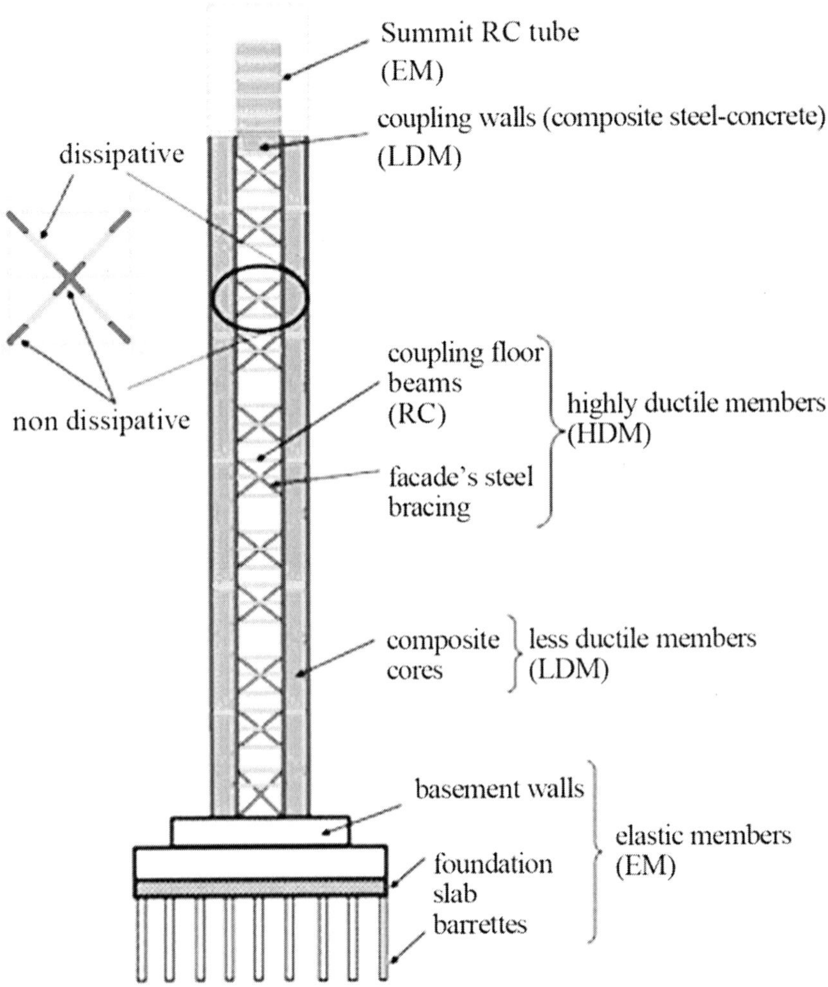

Figure 10. Classification of the structure members according to their designed seismic behavior.

5. ANALYSIS

To achieve and check out the aseismic design concept according to Figures 9 and 10 several analytical approaches have been used: modal analysis, push-over analysis and design capacity method. Seismic actions parallel to both the façades and to the floor diagonals have been considered.

The building weight has been either increased by 10% [3] or decreased by 20% [4] depending on which case was non-beneficial.

5.1. Modal Analysis

An elastic 3D-model of the entire structure from the foundation slab up to the very top of the building has been used. It has been assumed that the model is built-in at the base. Additional to each seismic action a simultaneous 30%—component in the perpendicular direction has also been considered [3].

The modal analyses have been used 1) to check out the compliance with the limit imposed by [3] for the elastic relative story displacement (1% of the story height), and 2) to proportion all structural members which could undergo plastic deformation during the design earthquake. To account for the plastic seismic response of the structure the design elastic seismic forces have been reduced with a "behavior" factor q equal to 3.6. The proportioning refers primarily to 1) the steel sections of the dissipative parts of the facade bracings, and 2) the longitudinal reinforcement of both ends of the coupling floor beams as well as of the cross-sections at the core bottoms.

5.2. Push-Over Analysis

The non-linear 3-D model used within the push-over analyses assumes that the tower is built-in at the ground level and made only of bars. The model has considered the geometrical effect of the box-type cores. The vertical distribution of the seismic loads has been assumed to be either rectangular or triangular [3].

The plastic behavior of the bars is modeled by means of bilinear relationships force—plastic deformation as depicted in Figure 11.

The bilinear relationships bending moments—end rotations for the RC bars (i.e. the floor coupling beams and the bottom of cores) were deducted by means of the constitutive relationships of concrete and reinforcement recommended in [5] with the mean values of both strengths and strains recommended within the Annex A. Both the confining effect of the transverse reinforcement and the influence of the axial forces have been taken into consideration. The reinforcement bars in the potential yielding zones have been designed with steel grade S500C in order to ensure the necessary ductility [5].

The bilinear relationships axial force—axial displacement for the steel bars of the façade bracing correspond to the values given in Table 1. Special requirements were imposed on the production of S235 to make sure that the actual yield stress complies with the assumed value. The façade bracings are designed with welded sturdy H-profiles. Their slenderness is less than 25 and the cross-sections correspond to class 1 according to [6].

On this account both tensile and compressive diagonals have been considered active. However, to account for the second order effects the assumed axial forces and ultimate deformations of compressive diagonals where reduced to 80% and, respectively, 75 % of the corresponding values of tensile diagonals.

For the calculation of the ultimate axial deformations of the façade bracings only the "dissipative" parts of the diagonals which are designed with S235 steel grade have been considered (see detail in Figure 10). Their length is 4000 mm at the 1st bracing and 3400 mm above.

The results of the push-over analyses are depicted in Figure 12.

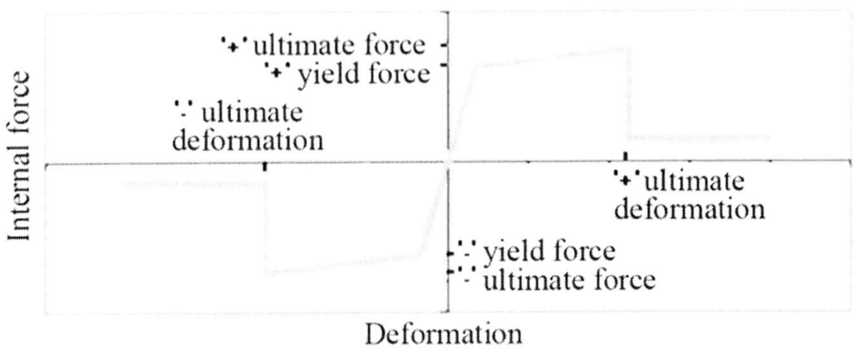

Figure 11. Bilinear force—deformation relationship used within the push-over analyses

Table 1. Values used to define the constitutive relationship of steel plates with thickness up to 40 mm.

Steel grade	Mean yield strength [MPa]	Mean ultimate strength [MPa]	Ultimate strain [%]
S235	245	307	3
S355	416	598	-

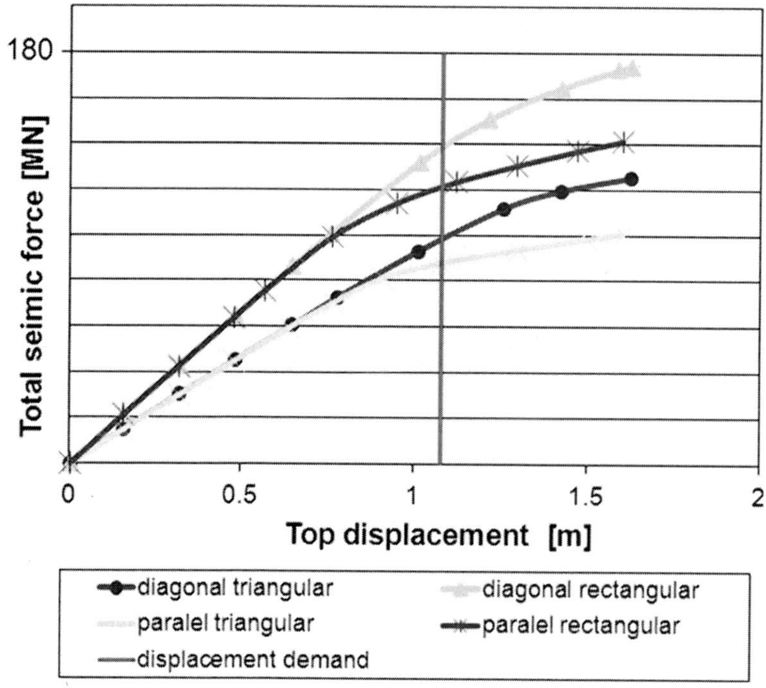

Figure 12. Capacity curves resulted from push-over analyses carried out for different directions of the seismic action and for different distributions of the seismic loads.

The red line corresponds to the displacement demand in Figure 9 and according to [3] is called "target displacement" (TD). TD has been evaluated according to the Annex B to [3] for the design spectrum from Figure 2 and resulted equal to 1.44 × 75 cm = 1.08 m, where 75 cm is the maximum displacement of the nonlinear equivalent system with a single degree-of-freedom (approximately the same for both seismic directions) and 1.44 is a factor accounting for the effect of multiple degreesof-freedom. The push-over analyses have been carried out up to a displacement 1.5 times higher than TD.

Figure 12 shows that the structure is expected to respond to the design earthquake without any significant damage and that the maximum base shear forces induced by the seism will range between 100 MN and 170 MN, depending on the direction of the seismic action and on the seismic load distribution over the building height. As expected the constant distribution yields higher seismic loads.

For comparison the corresponding values according to the modal analyses are 64 MN for the base shear force, practically independent of the earthquake direction, and 48 cm and 34 cm for the elastic top displacement when the seismic action is parallel to the façade and, respectively, to the floor diagonal.

A rough idea on the influence of the plastic behavior on the seismic response can be obtained by comparing the results of the two analyses. The ratio between the maximum seismic responses according to the push-over analysis and according to the modal analysis varies between 1.5 and 2.2 in terms of both forces and displacements. These values give a hint of the magnitude of the actual behavior factor q.

The push-over analyses yield some important detailed information on the plastic seismic response of the HDand LD-members from Figure 10, which are crucial for the energy dissipation. They are to be discussed hereafter only for the triangular vertical distribution of the seismic loads as this case is more plausible for the minaret. As expected for a mast-like tower with braced corner cores, the magnitudes of the plastic deformations of the facade bracings and of the coupling floor beams become larger when the seismic action is parallel to the façade, whereas the core dimensioning is associated with the seismic action along the floor diagonal.

The behavior of the façade bracings is depicted in Figure 13, which shows all 11 façade bracings over the minaret height as well as the steel profiles within the walls coupling the cores at their top. The given steps of the top displacement correspond to the last 5 cross— marks on the blue line in Figure 12. The points depicted "pink" denote the situations when the yielding is initialized; the points depicted "blue" the situations when the plastic deformations are smaller than 50% of the ultimate limit. As expected the compressive diagonals yield first and more than the tensile ones. At TD the largest plastic deformation occurs in the 4th bracing and reaches some 6% of the ultimate deformation. At 1.5 times TD the largest plastic deformations occur in the 6th bracing from bottom and amount to some 30% of the ultimate value in the compressive diagonal and, respectively, 20% in the tensile one. The 1st bracing and the steel bracing within the coupling walls at the top of the cores remain elastic even at 1.5 times TD.

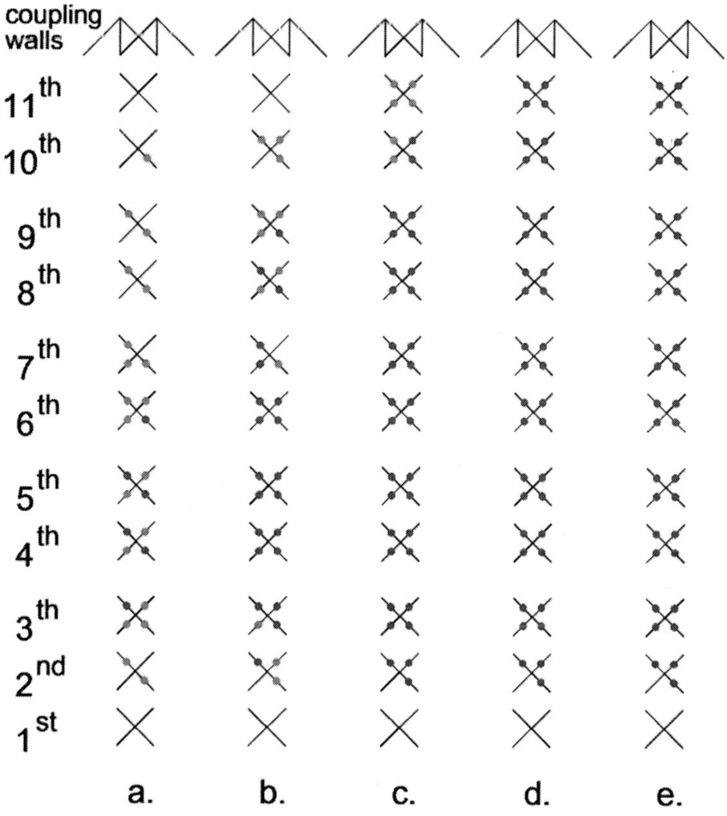

Figure 13. The degree of yielding of the façade bracings resulted from the push-over analysis carried out for the seismic action parallel to the façade with a triangular vertical distribution of the seismic loads. The situations depicted correspond to top displacements which are 85% (a), 100% (b), 120% (c), 140% (d) and, respectively, 150% of the TDvalue.

Concerning the plastic behavior of the coupling floor beams the results of the push-over analyses indicate that:

- At TD all beams of the floors situated between the 1st and the 9th bracing in Figure 13 yield, whereas the maximum plastic hinge rotation at the upper side amounts to 22% of the ultimate value and that at the lower side to 3% of its ultimate value.
- At 1.5 times TD the beam yielding extends to all floors up to the coupling walls at the top of the cores. The maximum plastic deformations occur within the floors between the tops of 3rd and the 8th bracings in Figure 13 and amount to 43% (at upper side) and 12% (at the lower side) of their respective ultimate values.
- Concerning the plastic behavior of the cores at minaret base the push-over analysis yields the following results
- At TD no plastic deformation has occurred.
- The first plastic hinge occurs within the tensile core when the top displacement amounts to a value 1.4 times higher as TD.
- At 1.5 times TD the maximum plastic rotation at the bottom of the tensile core amounts to 1% of its ultimate value while all other cores behave still elastically.

6. STRUCTURAL MEMBERS

The main structural members of the minaret are either RC elements (basement walls, all floor elements and the summit tube) or steel elements (façade bracings, substructure of the summit envelope) or composite elements (the cores and the coupling walls at the core tops). Some fundamental details concerning the proportioning of the bracings, of the cores and of the coupling floor beams will be presented hereafter.

When using the Euro Codes the proportioning of the structural elements must fulfill the requirement

$$E_d \leq R_d \tag{1}$$

E_d denotes the design internal forces and R_d the corresponding resistances. For the aseismic design the R_d values are determined by means of the material design strengths, which result from dividing the characteristic strengths by the factors $\gamma_c = 1.2$ for concrete and $\gamma_s = 1$ for steel.

In the case of a HD-member the E_d value results on the basis of the modal analysis. After their proportioning and detailing the actual resistance R_{eff} of the element can be calculated by using all existing cross-section components and more realistic material strengths as the design values, e.g. the mean values.

In the case of LDor E-members the E_d value results on the basis of push-over analyses by considering a top displacement equal to TD or more. The higher the top displacement considered, the safer is the proportioning of these elements.

In the special case of possible fragile collapses (e.g. bolted connections of steel members or shear resistance of RC elements required to behave ductile) the E_d values results by means of the "capacity design" method.

6.1. Steel Bracing

To the steel bracing belong not only the facade diagonals but also the steel profiles embedded within the core external walls (see Figure 4). The axial forces in the façade diagonals are given by their real resistance R_{eff}. On the basis of these forces the associated axial forces within the other steel bracing members can be determined by simply using equilibrium conditions. For the vertical steel members situated in the three external corners of the cores these axial forces represent only a local modification ΔN of the global axial forces N_{glob} resulting from the seismic response of the structure (Figure 14). Indeed the vertical steel members reinforce the core cross section and the N_{glob} forces are associated with the sectional forces N and M in Figure 6. Due to the stud connectors the local modifications ΔN are transferred to the other components of the core wall, i.e. to concrete and longitudinal reinforcement bars, so that the effect of ΔN disappears at a certain distance beneath the corresponding node and is already included within the global force N_{glob}. It has been assumed by the proportioning of the strut connectors on the vertical steel profiles that the transfer of the ΔN takes place over a length of a story height. When the vertical steel member is elastically stressed, ΔN is transferred both above and beneath the corresponding truss node. When the vertical steel member yields, ΔN is transferred only above the corresponding truss node, i.e. there where an unloading takes place.

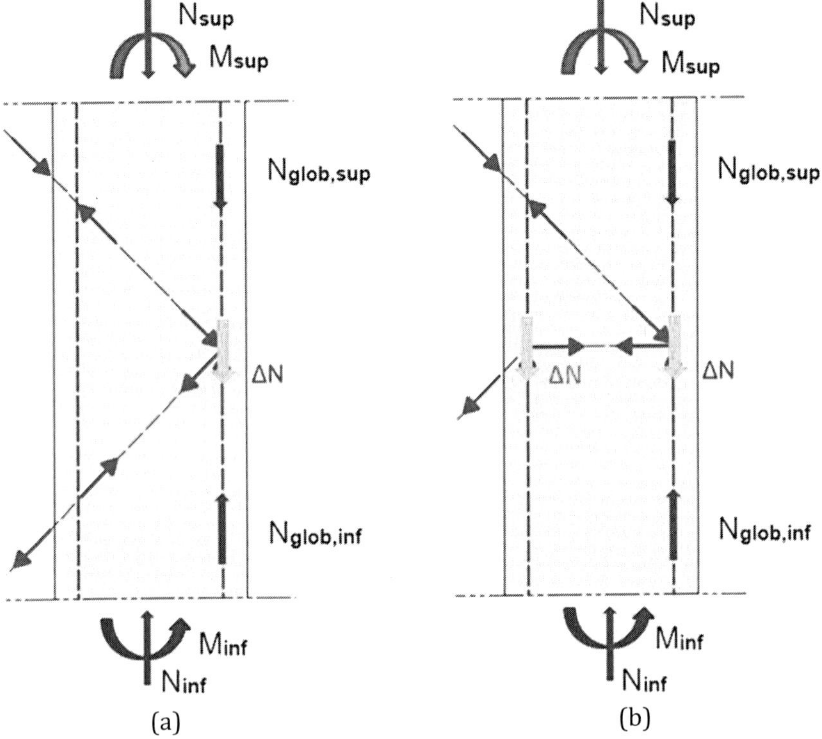

Figure 14. Superposition of local and global axial forces of the vertical steel members embedded in the core external walls: (a) over the height of a sky foyer, and (b) over the height of a story without façade diagonals.

Typical details of the steel bracing in elevation and in plan are depicted in Figures 15 and 16, respectively. The exterior vertical member is embedded at the junction of both external core walls. Its cross-section is made of two identical parts placed within each of the two external walls. The two parts are welded together at every node of the spatial truss (Figure 17). The interior vertical member is embedded in the external core wall at its junction with the internal core wall. To reduce the slenderness of the façade diagonals the central node of the X-cross joint is hold horizontally in the plane of the corresponding RC floor.

All connections to be executed on the construction site are bolted. The positions of all these connections were dictated by geometrical, transportation and erection criteria.

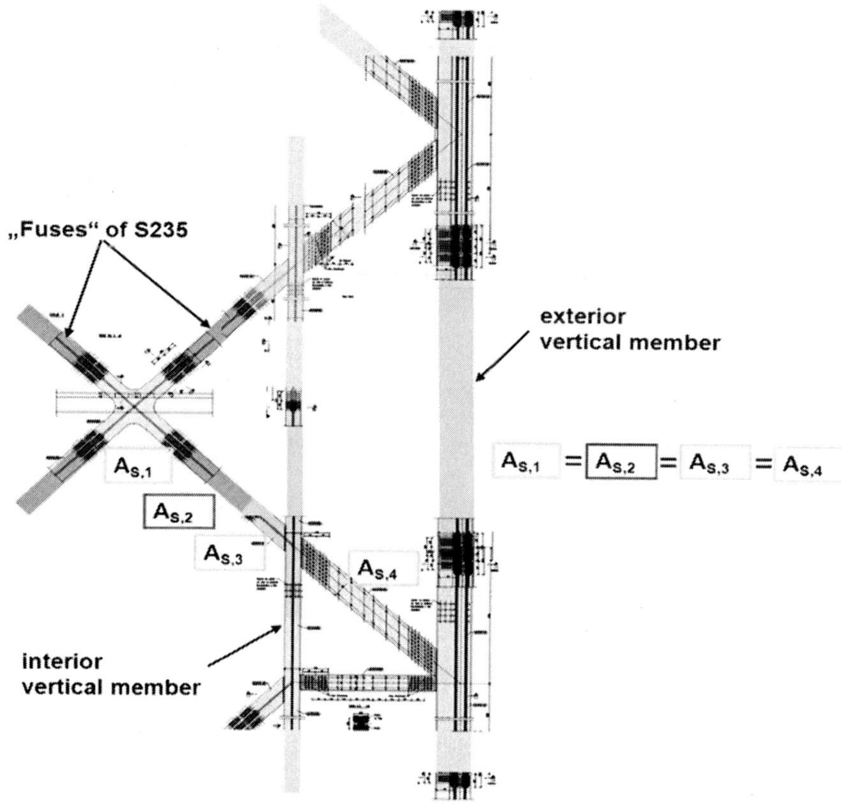

Figure 15. Typical elevation of the steel bracing.

Basically the design of the cross section areas and of the connections in Figure 15 follows the concept already described in conjunction with Figure 10.

- The parts of the façade diagonals with the cross section area $A_{s,2}$ made of highly ductile steel grade S235 act as "fuses" for the entire steel bracing.
- The other parts of the facade diagonals and the embedded diagonals have the same cross-section areas but are made of steel grade S355. The ratio between the yield strengths of the two steel grades is equal to $355/235 \approx 1.5$, i.e. large enough to ensure that the S355 parts remain elastic.

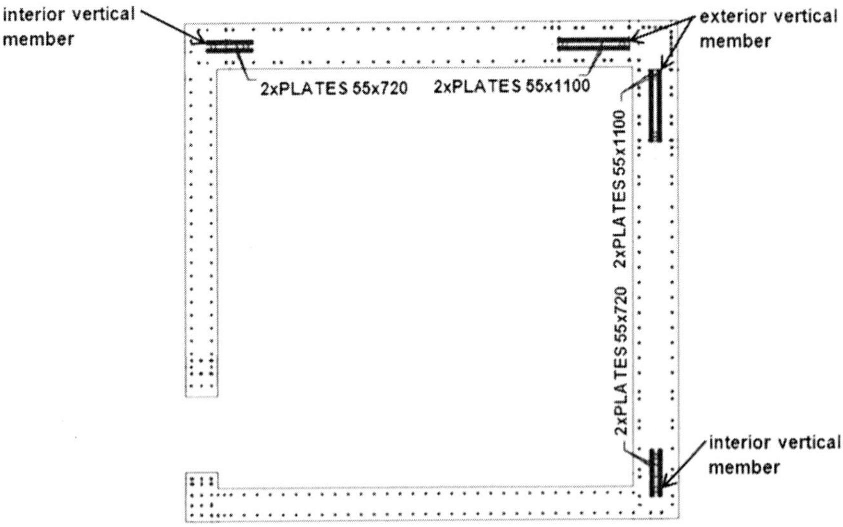

Figure 16. Core's composite cross section. The section depicted here corresponds to the middle part of the core height.

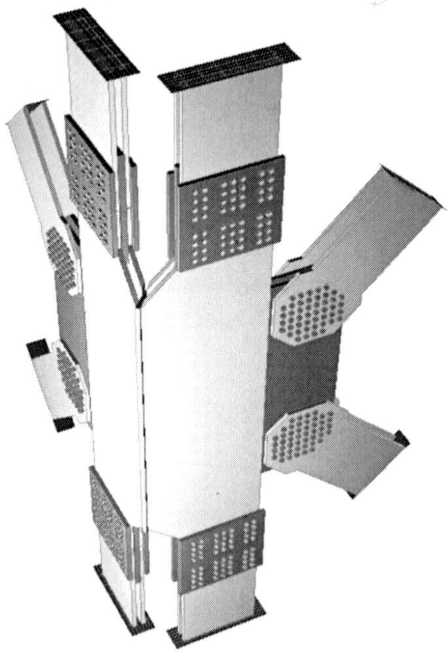

Figure 17. Detail of the corner joint of the spatial steel truss.

- The bolted and fillet weld non dissipative connections are also designed with a sufficiently safe overstrength relative to the steel profiles. The resistance R_{bd} of the bolted shear connections fulfills the relation [3]

$$R_{bd} \geq 1.2 \times 1.25 \times 1.1 \times R_{sd} \tag{2}$$

R_{sd} denotes the resistance of the connected dissipative part and is calculated with the cross section area $A_{s,2}$ and with the design yield stress $f_{sy,d}$ of S235. The factor 1.1 accounts for the possibility that the real value of $f_{sy,d}$ may be larger than considered, the factor 1.25 accounts for the necessary overstrength of a non dissipative connection and the factor 1.2 accounts for the necessary overstrength of shear resistance relative to the design bearing resistance. The resistance R_{wd} of fillet weld connections fulfills the relation [3]

$$R_{wd} \geq 1.25 \times 1.1 \times E_{wd} \tag{3}$$

The acting force E_{wd} is associated to R_{sd} and equal to the local force ΔN in Figure 14.

- The non dissipative connections made by means of full penetration butt welds are deemed to satisfy the overstrength criterion [3].

The two parts of the exterior vertical member in Figure 17 can receive different local forces ΔN depending on the earthquake direction (Figures 18 and 19). In the case "c" the two parts are similarly loaded, so that their connection is not stressed. On the contrary in the case "b" the connection is mostly stressed. The existing eccentricity between the forces ΔN yields a local rotational moment within the core external wall, which additionally stresses the strut connectors and which has to be considered by the proportioning of the horizontal reinforcement of the core external walls.

6.2. Coupling Floor RC Beams

During the design earthquake plastic hinges occur in both directions at both ends of the main beams of the RC floors (Figure 3). Additionally to the flexure produced by seismic action and gravity loads the outer main beams are also subjected to permanent tensile forces. Indeed the steel diagonals are compressed by the structure weight, push the adjacent cores apart of each other and therefore tension the floor beams (Figure 20(a)). The imposed compressive force D and subsequently the imposed tensile force N_B increase with the relative vertical shortening Δv of the cores and decrease with the elongation u of the RC beams. The creep of concrete core acts negatively while the cracking of the floor beams is beneficial. The final N_B corresponds to (3) in Figure 20(b), where $\varepsilon_v = \Delta v/H$ denotes the vertical strain associated with the relative vertical shortening Δv and the factor A is equal with $2/3 \cdot E_s \cdot A_{sD} \cdot \sin^2\alpha \cdot \cos\alpha$. Following notations are used: E_s = steel modulus of elasticity, A_s = total sectional area of the beam's longitudinal reinforcement, A_{sD} = the sectional area of the steel diagonal, ψ = parameter accounting for the beneficial effect of the tensile concrete on the actual beam elongation.

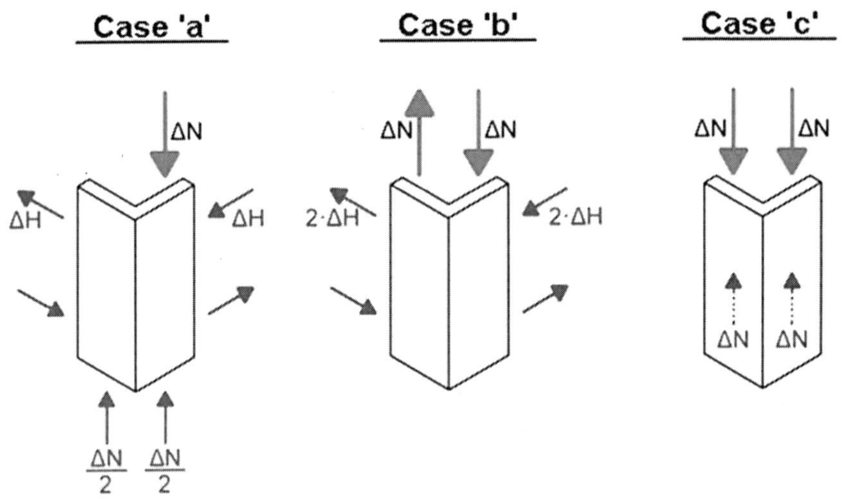

Figure 18. Possible local loading of the exterior vertical member dependent on the earthquake direction.

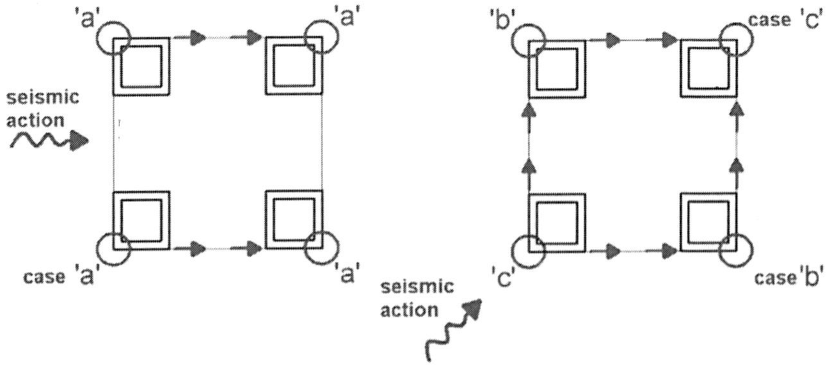

Figure 19. Correspondence between the cases depicted in Figure 18 and the earthquake direction.

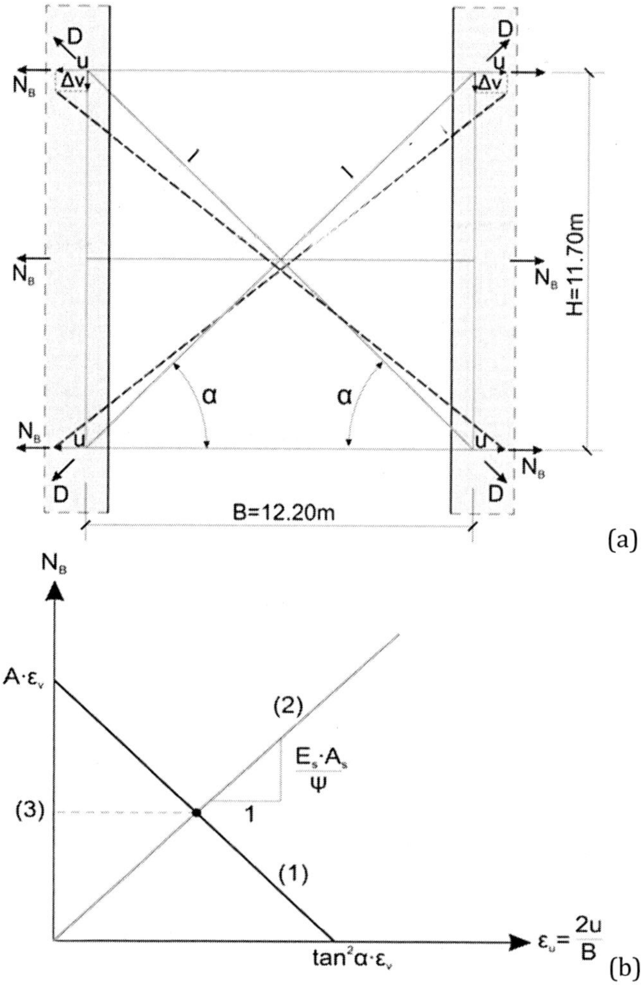

Figure 20. Effect of the core vertical shortening on the cracking of outer floor beam.

The straight line (1) in Figure 20(b) describes the effect of Δv on N_B and the straight line (2) corresponds to the beam elongation.

There is a correspondence between the final N_B value and the calculated crack opening. It has been attested that the crack openings comply with the limit of 0.3 mm imposed by [5] for indoor RC elements.

6.3. Composite Cores

The cores have a composite cross-section (Figure 16). The concrete has the class C50/60, the steel used for the reinforcement bars has the grade S500 and the embedded steel members are made of steel grade S355.

The necessary vertical reinforcement arose from the bending of the tower in a diagonal direction, whereas the necessary horizontal reinforcement arose from the bending of the tower in a direction parallel to façade. The results of the push-over analyses for the top displacement equal to the TD—value have been used. However the proportioning has been performed by means of bending moment envelopes of the results of both modal and pushover analyses, as the push-over analysis did not capture the influence of higher vibration modes. A typical example is given in Figure 21.

The vertical reinforcement at the bottom of the minaret has been finally decided by the internal forces of the tensile core. The safety margins of the both most stressed core cross-sections are outlined in Figure 22 by means of the difference between the acting bending moments M_{Ed} and the resistant moments M_{Rd}. The corresponding axial forces N are 265 MN (tension) and, respectively, 609 MN (compression).

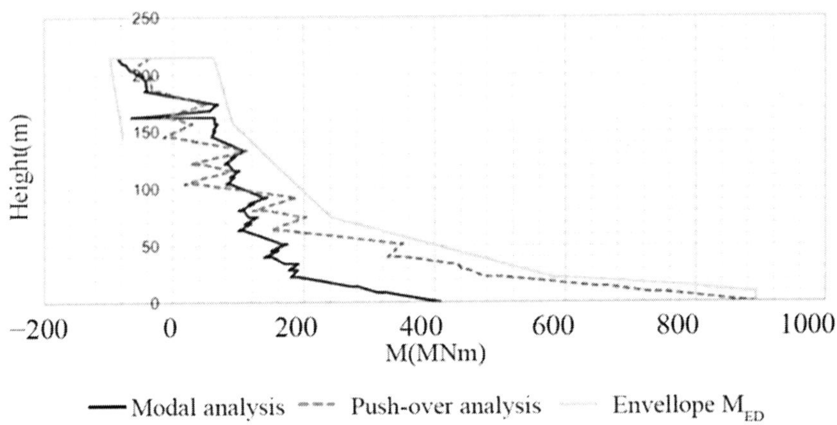

Figure 21. Bending moments of the compressive core by an earthquake along the floor diagonal.

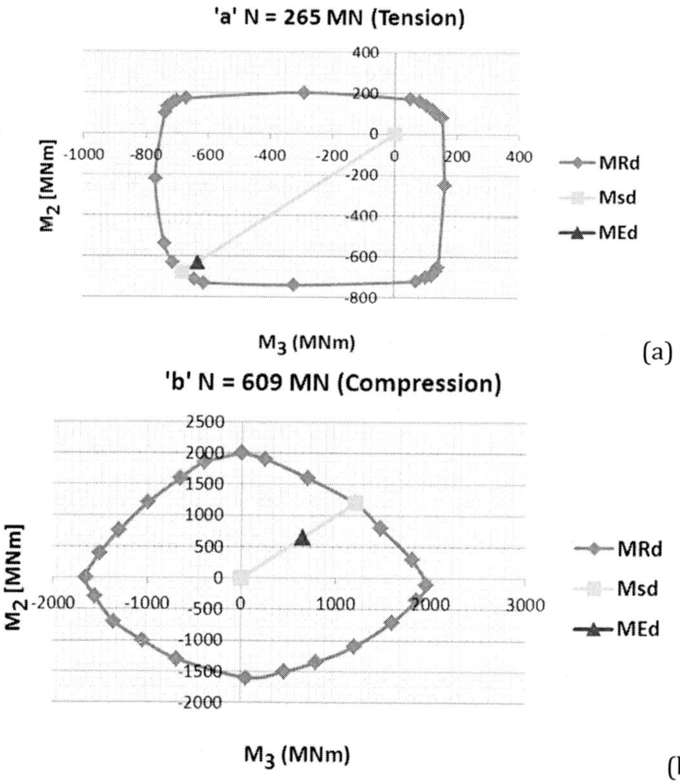

Figure 22. Relationship between the acting design bending moments M_{Ed} (red point) and the resistant moments M_{Rd} (blue interaction diagram) for the tensile (a) and the compressive (b) cores at the minaret base when the design earthquake acts along the floor diagonal.

The proportioning of the horizontal reinforcement is based on the envelope of the tower shear forces depicted in Figure 23. It has the form prescribed by [3]. The assumed value $V_{Ed,base}$ corresponds to the rectangular distribution of the seismic forces over the tower height and to a top displacement 1.5 times higher than TD (Figure 12). This aimed at ensuring a higher resistance to shear than to bending. The ratio between the total seismic forces considered for the two situations is 140 MN/100 MN = 1.4.

The required horizontal reinforcement of the core walls arises from the design shear forces of the walls. To obtain these, the floor shear force from Figure 23 had to be distributed first to the cores and afterwards to the core walls resisting shear. The distribution factors were chosen by means of the elastic model used for the modal analyses. Both the earthquake direction parallel to façade as well the direction along the floor diagonal were analyzed.

Concerning the distribution of V_{Ed} between the four cores it has been found that 1) it is practically equal, i.e. $V_{Ed,core} \approx 0.25 \times V_{Ed}$, when "parallel" earthquake and that (ii) each of the two cores situated on the diagonal along which the "diagonal" earthquake acts overtakes only some 20% of the floor shear force, so that the relevant design value became $V_{Ed,core} \approx 0.3 \times V_{Ed}$.

Concerning the distribution of $V_{Ed,core}$ between the resisting walls it has been found that it can be considered proportional to the wall thickness if a corrective factor is used for the external wall. The value of 1.15 has been found conservative in the case of the "parallel" earthquake, which has proved to be decisive for the required horizontal reinforcement. The required horizontal reinforcement A_{sw}/s (cm²/m) has been determined by means of Equation (1) with E_d taken equal to the value of $V_{Ed,wall}$ as yielded by the described distribution of V_{Ed} and R_d taken equal to the value

$$R_{wd,wall} = A_{sw}/s \times 0.85 \times h_w \times f_{sd}/1.15 \tag{4}$$

h_w denotes the length of the wall cross section.

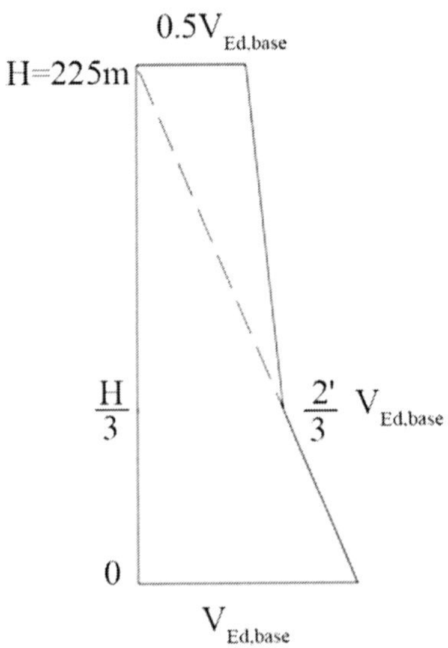

Figure 23. Design shear forces V_{Ed}.

The required horizontal reinforcement has been provided over the entire story height neglecting therefore the major beneficial contribution of the steel bracing. Indeed over the height of the stories with braced façade the steel diagonals overtake the most part of V_{Ed} whereas over the height of the other stories the steel diagonals embedded in the external walls have an important contribution to $R_{wd,wall}$. The additionally provided shear resistance accounts for the negative effects of local horizontal tensile stresses (e.g. see the discussion made in conjunction with the case "b" in Figure 18) and prevents the cracking caused by the composite action.

7. FOUNDATION SYSTEM

To this system belongs the rigid box-type enlargement of the minaret foot over the height of the two underground levels, the foundation slab and the "barrettes". All structural members are of cast-in RC with concrete C50/60 (the basement walls) and C30/37 (floors, the foundation slab and the "barrettes") and with reinforcement bars of steel grade S500. In accordance with the general design philosophy of the minaret, the foundation system has a higher seismic resistance as the tower itself. Moreover the resistance is increased gradually from the tower base to the foundation soil.

The design of the underground levels and of the foundation slab has been performed by means of a 3D-elastic model loaded with forces at the core bottoms and supported by elastic springs at the centers of "barrettes" and concomitantly by the subgrade soil.

The walls were modeled with membrane finite elements and the slabs (floors and foundation) with slab— membrane elements.

The forces at the core bottoms correspond to the results of the push-over analyses for a top displacement equal to 1.5 times TD. The axial forces and the biaxial bending moments at the base of each core have been considered by means of equivalent vertical forces applied at the corners of each core. The base shear forces have been given at the geometric center of each core. Different directions of the seismic action have been considered.

Each supporting point resists 3D displacements, i.e. two horizontal and one vertical. The corresponding spring stiffness values have resulted from the analyses of the soil-structure interaction reported below. The RC walls have to resist huge in-plane forces caused both by the cantilever action associated with the enlargement of the tower foot over the basement height and by the large seismic forces. These forces have required wall thicknesses up to 1.5 m and up to 8 layers of reinforcement grids.

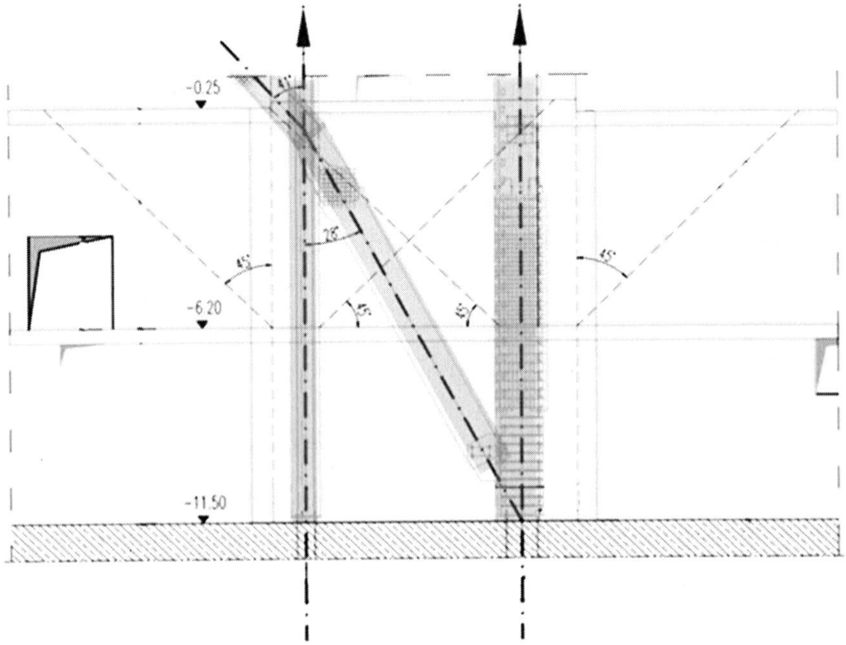

Figure 24. Anchorage of the tensile steel members at the tower bottom within the basement.

A special attention has to be paid to two "hanging" effects. The first corresponds to the anchorage of the tensile forces within the steel members at tower foot and the second to the anchorage of the tensile "barrettes", i.e. of those withstanding the foundation uplifting.

The first effect can occur either at the junction of the tensile façade diagonal with the interior vertical steel member or at the minaret corner when the vertical steel member is subjected to tension. The corresponding anchorage has been realized by embedding steel members over the entire basement height (Figure 24) and by providing sufficient vertical and horizontal reinforcement to spread the tensile forces and lead them into the foundation slab.

The second effect occurs at the border of the foundation slab at the junction with the "barrettes" which prevent the building's uplifting. Due to the magnitude of the "barrette" tensile forces (up to 37 MN) and to the cantilever caused by the eccentric peripheral wall (Figure 8) the foundation slab must resist very large shear forces.

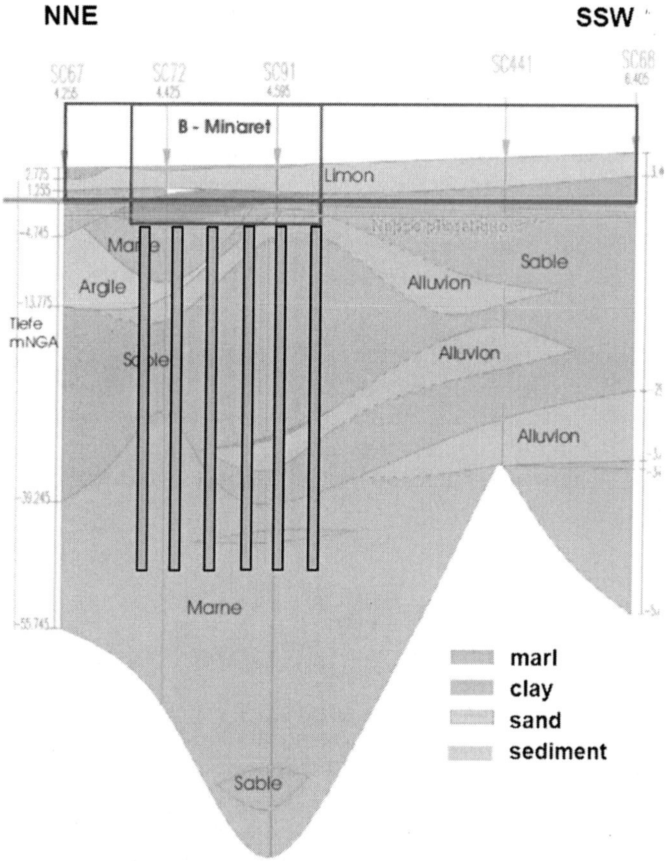

Figure 25. Geotechnical profile.

To realistically capture the soil-structure interaction an elastic 3D model with volumetric finite elements has been used. It consists of a 9 m thick foundation slab, of the "barrettes" and of the foundation soil. With the chosen thickness the modeled foundation slab has the same flexural stiffness as the entire box-type basement. The soil characteristics have been chosen according to the results of in situ soil tests carried out by the National Geotechnical Laboratory at Algiers (LCTP). A vertical section through the foundation soil is given in Figure 25. The barrettes traverse a thick sand layer and penetrate into the marl layer.

The forces acting at the core bottoms have been taken larger than the ones used to design the basement and foundation slab. Thus the vertical forces at the core corners associated with the seismic action have been amplified by the factor 1.05 whereas the horizontal forces by the factor 1.15. The higher amplification of the horizontal forces should cover the possibility that the total seismic force could act lower as assumed. Eight directions of the seismic action have been considered, i.e. for both directions of each of the following seismic actions: parallel to the two adjacent façades and to the two floor diagonals. That was necessary in order to capture the influence of the different depths of the soil layers.

The modeling has shown that the soil is capable of carrying the forces induced and, as same time, has provided the internal forces and the 3D settlements at the "barrette" tops. These results have been used 1) to proportion the required barrette reinforcement, and 2) to evaluate the springs stiffness needed for the modeling of the basement and of the foundation slab.

8. CONCLUDING REMARKS

The design of the lateral stiffness and resistance of the minaret has been decisively affected by the expected seismic action. The wind action has been investigated by means of wind-tunnel tests carried out with 1/400 models. Both rigid and deformable models have been investigated. They have shown that the expected base shear force induced by the local wind conditions will reach about 25 MN, which is several times smaller than the design base shear force induced by the design earthquake. Push-over analyses indicated values of 100 MN - 140 MN for the latter forces, depending on the earthquake direction and on the distribution of the seismic forces over the tower height. According to these analyses the expected maximum displacement at the tower's top amounts to 1.08 m and is associated with a non-linear seismic response.

The aseismic design has been performed on the basis of the performance criteria recommended by [3], which aims at ensuring optimal energy dissipation during the design earthquake and at protecting simultaneously the structural members which are to respond quasi elastically. The requirement for elastic response arises either for those structural members and connections whose collapse may be fragile or for components of the foundation system whose behavior depends on the soil response.

Consequently the member's strength has been gradually increased from the highly ductile members over the less ductile ones up to the members which should respond elastically. The first category includes the façade steel bracing and the RC main floor beams. The second includes the bottom region of the tower cores and the coupling structural walls at the top of the cores. The latter category includes the rest of the core height and the members of the foundation system, i.e. the box-type basement, the foundation slab and the "barrettes".

The push-over analyses confirmed the design philosophy. Plastic deformations were registered only within the zones with ductile behavior and their maximal values have had a large safety margin to the ultimate ones, even for top displacements larger than 1.5 times TD.

Due to the strong seismic conditions, to the tower slenderness and to the discontinuous façade bracing, the structural design of the minaret has faced some challenging aspects. The paper has outlined the most important of them.

9. ACKNOWLEDGEMENTS

We do thank all the colleagues who participated at the design and detailing of the minaret tower. The contributions of K. Golonka, M. Friedrich and M. Neacsu with his team deserve a special acknowledgement.

REFERENCES

1. J. Engel, "Djamaâ el Djazair. The New Mosque at Algiers," Symposium '25th Anniversary of Krebs & Kiefer Karlsruhe', 2013, (unpublished).

2. D. Constantinescu and J. Akkermann, "Auslegung von Bauwerken gegen Erdbeben nach Eurocode 8. Anwendung am Beispiel internationaler Projekte," Proceedings of the Dresden Conference on Steel Constructions, Dresden, 24 March 2011, pp. 205-252.

3. Eurocode 8, "Design of Structures for Earthquake Resistance, Part 1: General Rules, Seismic Actions and Rules for Buildings," European Committee for Standardization, Brussels, 2004.

4. RPA 99, "Règles Parasismiques Algériennes," Centre National de Recherche Appliquée en Génie Parasismique, Algiers, 2003.

5. Eurocode 2, "Design of Concrete Structures, Part 1-1: General Rules and Rules for Buildings," European Committee for Standardization, Brussels, 2002.

6. Eurocode 3, "Design of Steel Structures, Part 1-1: General Rules and Rules for Buildings," European Committee for Standardization, Brussels, 2002.

4

CHAPTER

SATELLITE REMOTE SENSING IN SEISMOLOGY. A REVIEW

Andrew A. Tronin

ABSTRACT

A wide range of satellite methods is applied now in seismology. The first applications of satellite data for earthquake exploration were initiated in the '70s, when active faults were mapped on satellite images. It was a pure and simple extrapolation of airphoto geological interpretation methods into space. The modern embodiment of this method is alignment analysis. Time series of alignments on the Earth's surface are investigated before and after the earthquake. A further application of satellite data in seismology is related with geophysical methods. Electromagnetic methods have about the same long history of application for seismology. Stable statistical estimations of ionosphere-lithosphere relation were obtained based on satellite ionozonds. The most successful current project "DEMETER" shows impressive results. Satellite thermal infra-red data were applied for earthquake research in the next step. Numerous results have confirmed previous observations of thermal anomalies on the Earth's surface prior to earthquakes. A modern trend is the application of the outgoing long-wave radiation for earthquake research. In '80s a new technology—satellite radar interferometry—opened a new page. Spectacular pictures of co-seismic deformations were presented. Current researches are moving in the direction of pre-earthquake deformation detection. GPS technology is also widely used in seismology both for ionosphere sounding and for ground movement detection. Satellite gravimetry has demonstrated its first very impressive results on the example of the catastrophic Indonesian earthquake in 2004. Relatively new applications of remote sensing for seismology as atmospheric sounding, gas observations, and cloud analysis are considered as possible candidates for applications.

Keywords: remote sensing; seismology; earthquakes

1. INTRODUCTION

Remote sensing has been used for earthquake research from the '70s, with the first appearance of satellite images. First of all it was used in structural geological and geomorphological research. Active faults and structures were mapped on the base of satellite images. This method is very limited in time series analysis. There was no possibility to measure short term processes before and after the earthquake. It was simple an extrapolation of airphoto geological interpretation methods into space. The modern version of this method is active tectonic analysis with the application of alignment analysis. Time series of alignment distributions on the Earth's surface are investigated before and after an earthquake.

The current situation of remote sensing application for earthquake research indicates a few phenomena, related with earthquakes, particularly the Earth's surface deformation, surface temperature and humidity, atmosphere temperature and humidity, gas and aerosol content. Both horizontal and vertical deformations scaled from tens of centimeters to meters are recorded after the shock. Such deformations are recorded by the Interferometric Synthetic Aperture Radar (InSAR) technique with confidence. Pre-earthquake deformations are rather small, on the order of centimeters. A few cases of deformation mapping before the shock using satellite data are known at present time. Future developments lay in precision longwave SAR systems with medium spatial resolution and combined with the GPS technique. There are numerous observations of surface and near surface temperature increases of 3–5 °C prior to Earth crust earthquakes. Methods of earthquake prediction are developing using thermal infrared (TIR) surveys. Multiple evidence of gas and aerosol content changes before earthquakes are reported for ground observations. Satellite methods allow one to measure the concentrations of gases in atmosphere: $O3$, $CH4$, $CO2$, CO, $H2S$, $SO2$, HCl and aerosols. However the spatial resolution and sensitivity of modern systems restricts the application of satellite gas observation in seismology and the first promising results have been obtained only for ozone, aerosol and air humidity.

A few reviews have covered the field of satellite data applications for natural hazards [1,2,3,4]. Only some of them were concentrated on earthquake research [5,6,7]. A special issue of the Journal of Asian Earth Sciences entitled "Validation of Earthquake Precursors (by Satellite, Terrestrial and other Observations) VESTO" will be published at the near future.

2. DEFORMATIONS

One of the main directions of remote sensing application for seismology is deformation mapping. Surface deformations in seismic cycles can be divided into three phases: pre-seismic or inter-seismic, co-seismic and post-seismic ones. Co-seismic deformations are evaluated up to meters and tens of meters while pre-seismic movements amount to centimeters. Post-seismic deformations are also measured in centimeters, but subsequent landslides can increase deformations to meters. Most current research is focused on co-seismic and post-seismic (landslide) deformations.

2.1. Optical Sensors

The first application of satellite images in seismology was related to structural geology and geomorphology. Active faults and neotectonics were the research aims [8,9]. Epicenter zones of recent earthquakes were studied on the space images. Image interpretation depended on visual methods. The interpreter selected faults with sharp borders, shift of river valleys, etc. One example of such an application is shown on Figure 1.

Co-seismic and post-seismic (landslides) deformations are mapped by modern optic satellite systems. Typical example of high resolution satellite images application for surface deformation is presented on Figure 2. Change detection methods are widely used for landslide deformation mapping.

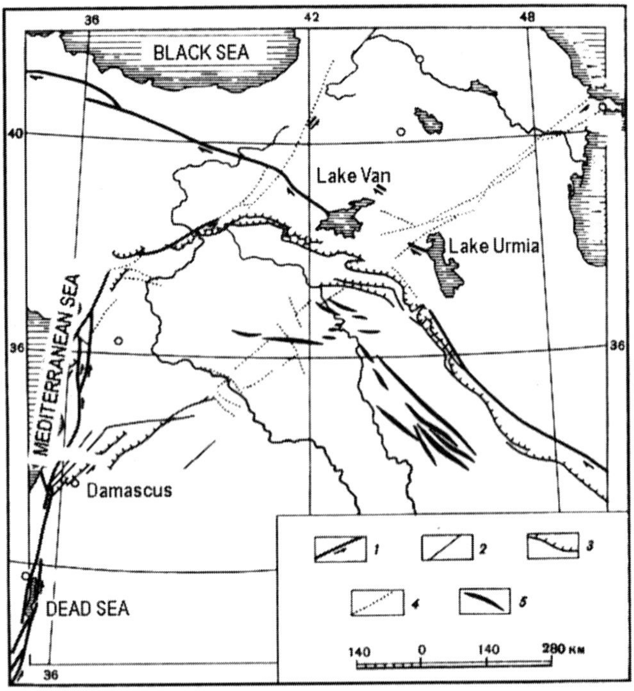

Figure 1. Main fault systems of North Arabian plate [10]:1—active faults; 2—late Cenozoic faults; 3—Cenozoic thrusts; 4—alignment on the base of satellite images; 5—anticlines.

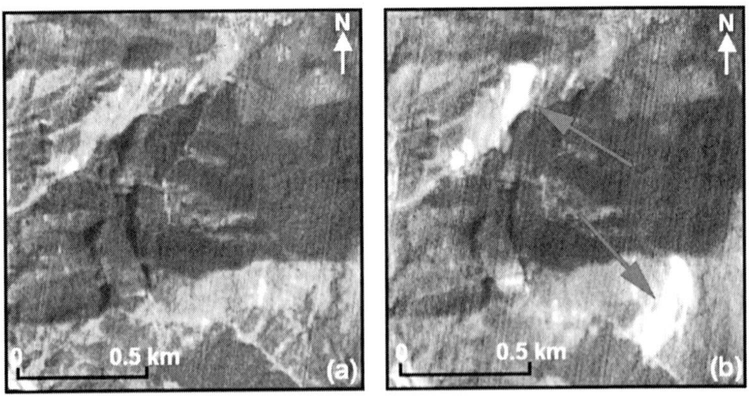

Figure 2. Surface deformations mapping for the Chamoli earthquake (M = 6.3, 29 March 1999 Himalaya, India). Change detection after the earthquake [11]. (a) IRS-1C-PAN pre-earthquake image of 26 March 1999. (b) IRS-1C-PAN post-earthquake image of 31 March 1999.

Deformation mapping with optical systems as a technique to measure horizontal movements was developed recently, after the InSAR deformation method (see Section 2.2). The technique is based on a sub-pixel correlation technique [12]. Mis-registration due to stereoscopic effects is compensated together with those due to the changing attitude of the satellite during image acquisition. Residual mis-registrations then reflect ground deformation. SPOT panchromatic images were applied to the Landers earthquake, 1992, USA [12]; the Izmit earthquake, 1999, Turkey [13], and the Chi-chi earthquake, 1999, in Taiwan [14]. The technique provides measurements with an accuracy of about 0.5 m. The displacement field obtained from this technique can be used to map co-seismic ground deformations and measure slips on the fault. InSAR technique provides precise measurements a few kilometers away from the faults but usually can not generate complete deformation map in the near-fault zone. Optical displacement method can supplement InSAR technique to cover this disadvantage of the InSAR technique.

Lineament analysis has become the new step in the application of optical data sensors in seismology. Time series of alignment distributions on the Earth's surface are investigated before and after an earthquake. Significant changes in alignment distributions (density, direction) were recorded before an earthquake [15].

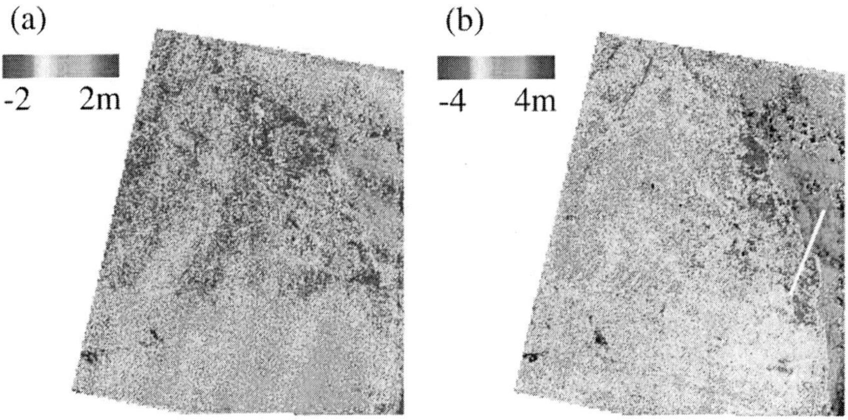

Figure 3. Horizontal ground displacement induced by the Landers earthquake measured from SPOT2 panchromatic images (pixel size 10 m). (a) East-West. (b) North-South components [12].

The application of optical methods for deformation mapping has limited use now due to cloud problems. It does not allow one to get long time image series for analysis. On the other hand, the physical basis of optical sensor applications for deformation mapping before the shock are not clear. The main trend of optical sensors in the nearest future will be related with post-seismic deformation mapping, especially in epicenter areas, with high resolution satellite systems.

2.2. InSAR

The InSAR technique is used to examine small-scale features in the deformation field associated with earthquakes. Satellite interferometry is based on multitemporal radar observations. InSAR is a method by which the phase differences of two or more SAR images are used to calculate the differences in range from two SAR antennae having slightly different viewing geometries to targets on the ground. As a result, displacements on the Earth's surface in range of centimeters and millimeters can be measured. The InSAR results show significant deformation signatures associated with faults, fractures and subsidences. The interferogram also clearly indicates surface deformation related to earthquakes.

The first application of satellite interferometry for earthquake research was demonstrated in the '90s by Massonnet et al. [16]. The well-known "butterfly" image (Figure 4) of the Landers earthquake (M = 7.3, 28 June 1992) was compiled on the base of pre-seismic image of April 24, 1992 and post-seismic scenes: August 7, 1992; 3 July, 1992 and June 18, 1993 [17]. The comparison of the observed interferometric picture and modeled scheme of earthquake deformations shows many similar details (Figure 5). Similar images have been obtained for all the significant earthquakes of the last decade: Neftegorsk, 1995, Russia [18], Bhuj, 2001, India [19]; Bam, 2003, Iran [20]; Sumatra (2004), Indonesia [21] etc. All these cases demonstrate co-seismic and post-seismic deformations.

First results with pre-seismic deformations recorded with InSAR technique were obtained recently. Probable pre-seismic deformation in the Tokai region, Japan (Figure 6) is indicated [22] on the multiyear InSAR data. The Tokai region is a well-known seismic gap located about 200 km southwest of Tokyo, at the conjunction of the Eurasian plate, the Philippine sea plate and the North-American plate. The vertical deformation recorded on the base of InSAR data coincides with ground leveling and GPS observation (Figure 10). The subsidence continues to this moment, but an earthquake has still not happened.

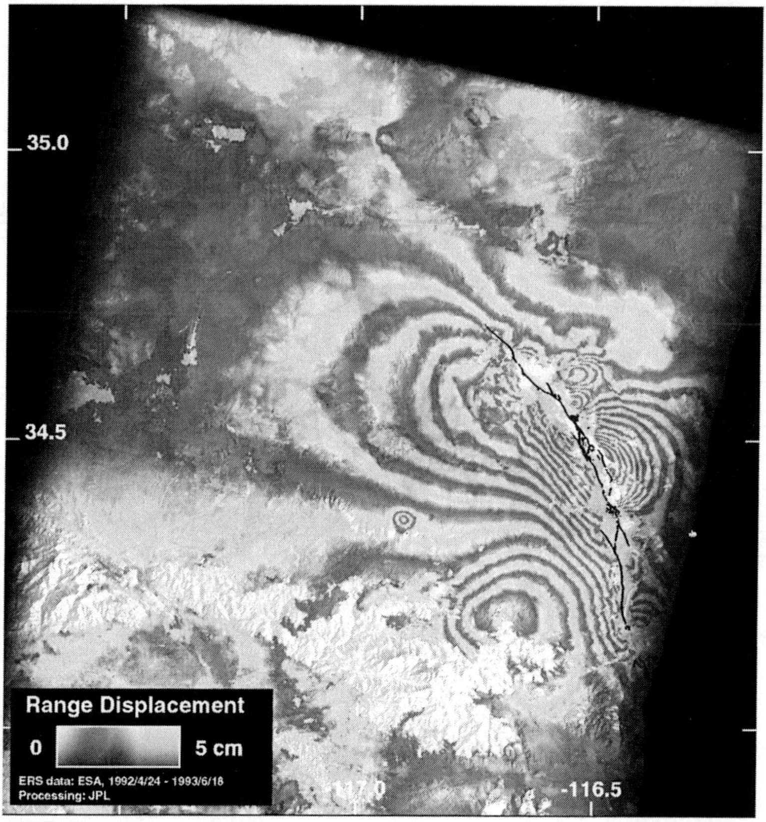

Figure 4. Observed co-seismic interferogram for the Landers earthquake (M = 7.3, 28 June 1992, CA, USA). One full color cycle represents here 5 cm of range displacement. Solid lines depict the fault geometry as mapped in the field. Gray areas are zones of low phase coherence that have been masked before phase unwrapping [17].

The following research has shown surface deformations three year before the Chi-chi earthquake (1999, Taiwan [23]). ERS-2 radar scenes were used to identify possible precursory surface deformation in the areas of the main fault before the Chi-Chi earthquake. It was found that surface deformation began at least three years before the Chi-Chi earthquake in the areas to the west of the northern segment of the fault where clear co-seismic surface deformation patterns were observed (Figure 7).

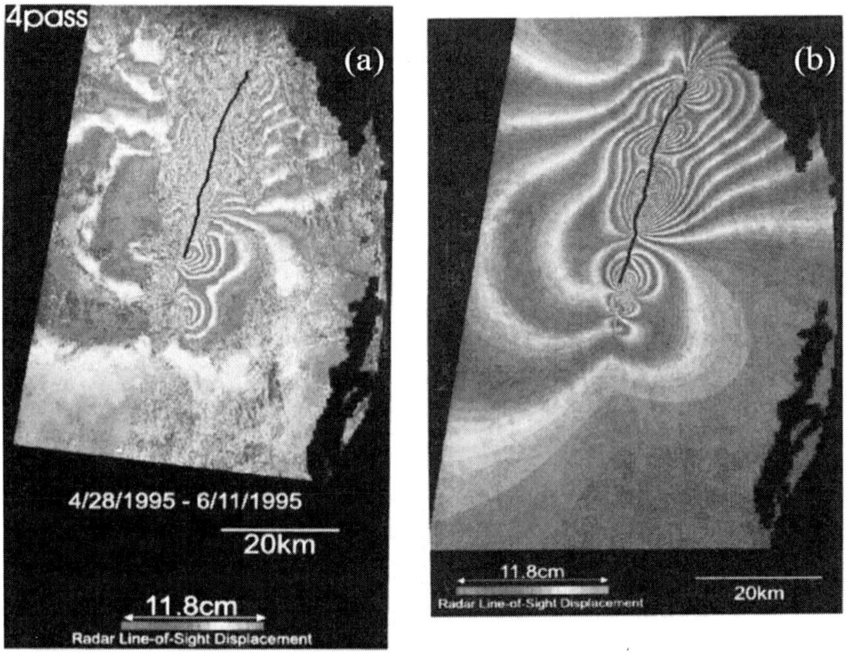

Figure 5. Ground deformations induced by the Neftegorsk earthquake (M = 7.6, 28 May 1995, Sakhalin, Russia). (a) Radar interferogram. (b) Deformation model prediction [18].

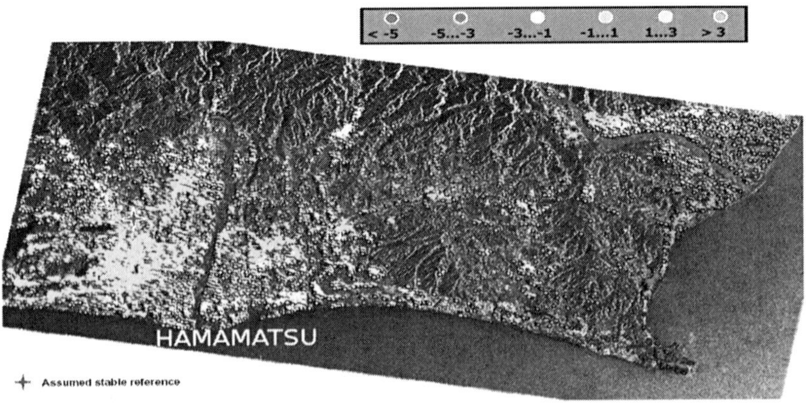

Figure 6. Average displacement rate in Tokai area, Japan, 1992–2000 years, vertical deformation, mm/year [22].

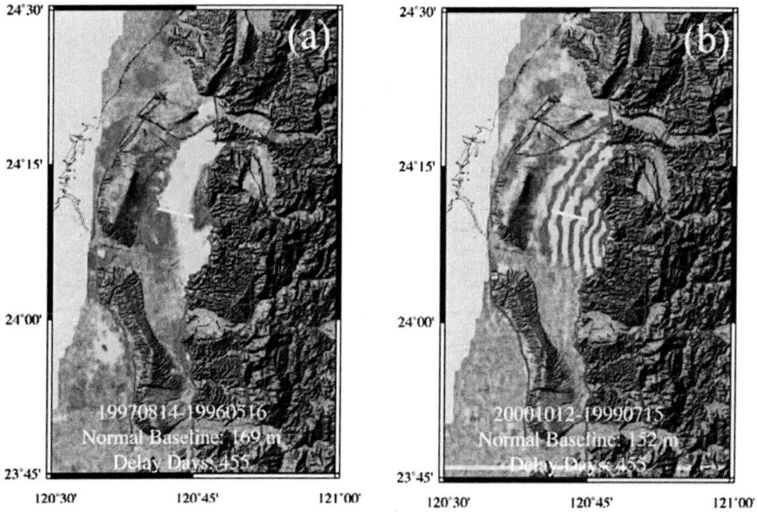

Figure 7. Pre-seismic and co-seismic ground deformations for the Chi-Chi earthquake (M = 7.6, 20 September 1999, Taiwan). (a) Pre-seismic deformations between 16 May 1996 and 14 August 1997. (b) Co-seismic ground deformations between 15 July 1999 and 12 October 2000 [23].

One of the other applications has described pre-seismic deformations related with the 2004 Parkfield earthquake [24]. Pre-seismic movements are clearly recorded for strike-slips like the San Andreas Fault Zone by InSAR. The relative horizontal speed of the slip is evaluated at about 2 cm/year. The spatial pattern of the pre-seismic InSAR deformation rate (Figure 8) indicates that strain distribution is not uniformly partitioned along the fault strike.

Post-seismic deformations also became the object of research. It was possible to separate co-seismic and post-seismic deformations for the Bam earthquake (Iran). A subsidence of about 3 cm was observed (Figure 9) in the area of the destructive fault three years after the main shock [25].

Figure 9. Post-seismic deformations for the Bam earthquake (M = 6.6, 26 December 2003, Iran). (a) Vertical displacement of the land surface south of Bam, during the three and a half years after the earthquake derived from analysis of radar images. The dark blue area sank a total of more than 3 cm (1.2 inches), revealing a zone of rock that was damaged during the earthquake and then healed afterwards. (b) False-color Landsat Thematic Mapper image taken on October 1, 1999 of the area of the earthquake rupture south of Bam. Red line shows location of the fault damage zone above the buried 2003 rupture. Vegetation in the city of Bam is green and stone-covered desert has various tones of gray [25].

The further enhancement of the InSAR technology will allow us to record very fine difference in surface displacement. The currently operated COSMO/SkyMed mission aims to provide daily observations, overcoming limited observational frequency, by using a constellation of four satellites. The current constellation consists of three medium-size satellites, each one equipped with a microwave high-resolution SAR operating in X-band (31 mm), having ~600 km single side access ground area, orbiting in a sun-synchronous orbit at ~620 km height over the Earth surface, with the capability to change attitude in order to acquire images at both right and left side of the satellite ground track. TERRASAR-X (wavelength-31 mm) radar also provides reliable information for surface deformation measurements.

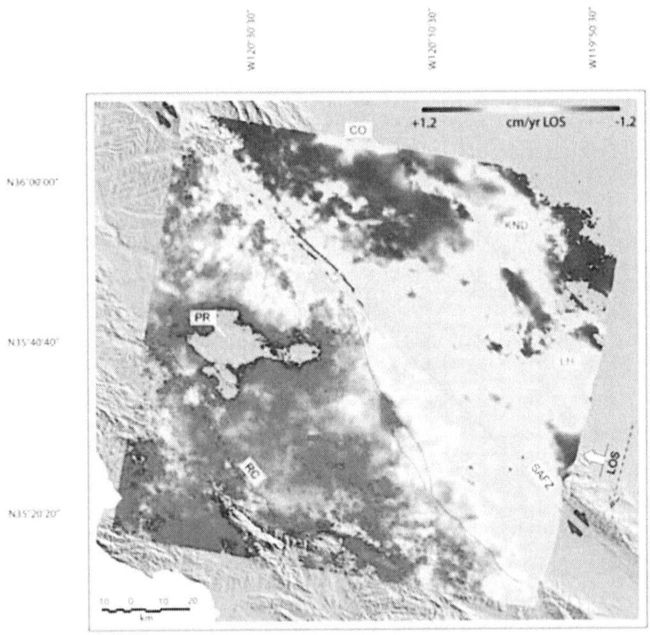

Figure 8. Pre-seismic strain buildup on the San Andreas Fault Zone at Parkfield. The map is composed by averaging 20 unwrapped ERS-1/2 interferograms in a stack. Each 20 × 20 m pixel is then a measure of the average in the line of sight direction velocity in the period 1992–2004 prior to the earthquake. The scale-bar values represent centimeters parallel to the fault [24].

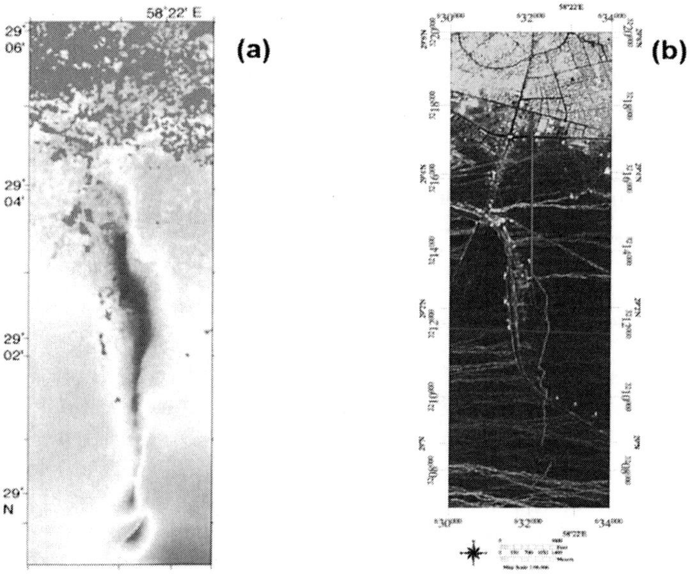

Figure 9. Post-seismic deformations for the Bam earthquake (M = 6.6, 26 December 2003, Iran). (a) Vertical displacement of the land surface south of Bam, during the three and a half years after the earthquake derived from analysis of radar images. The dark blue area sank a total of more than 3 cm (1.2 inches), revealing a zone of rock that was damaged during the earthquake and then healed afterwards. (b) False-color Landsat Thematic Mapper image taken on October 1, 1999 of the area of the earthquake rupture south of Bam. Red line shows location of the fault damage zone above the buried 2003 rupture. Vegetation in the city of Bam is green and stone-covered desert has various tones of gray [25].

Existing satellite InSAR instruments have X-band and C-band (a wavelength of 56.6 mm), offering high resolution, but they only provide reliable interferograms for coherent, non-vegetated surfaces. Data from the JERS-1 satellite demonstrated during its lifetime that L-band (236 mm wavelength) satellites propose reduced resolution but provide interferograms over a far greater range of surface cover types. Modern L-band (236 mm wavelength) SAR PALSAR onboard Japanese satellite ALOS supplies high quality data for interferometry. Unfortunately, this instrument is designed to test applications other than interferometry, so it provides only limited support for deformation analysis. Future missions like MOSART (Monitoring of Surface Deformation in Active Tectonic Zones) and TERRASAR-L are still not realized.

The InSAR technique opens new fields of remote sensing applications in seismology. InSAR methods are the best tool to study earthquake deformations at the moment after the shock. New systems will allow us to shift to pre-seismic deformation measurements and earthquake forecasting. Epicenter area deformations are still problem for InSAR due to coherence loss and permanent scatter destruction.

2.3. GPS

As GPS observations are not strictly a remote sensing application, only a few examples of their use for the study of surface deformations will be listed below. It was mentioned above that InSAR technique is more sensitive for vertical deformations, while the GPS method is capable of recording long period horizontal movements. Vertical ground deformations in Tokai area, Japan, revealed by InSAR (Figure 6) were accompanied by horizontal ones recorded (Figure 10) by GPS [26].

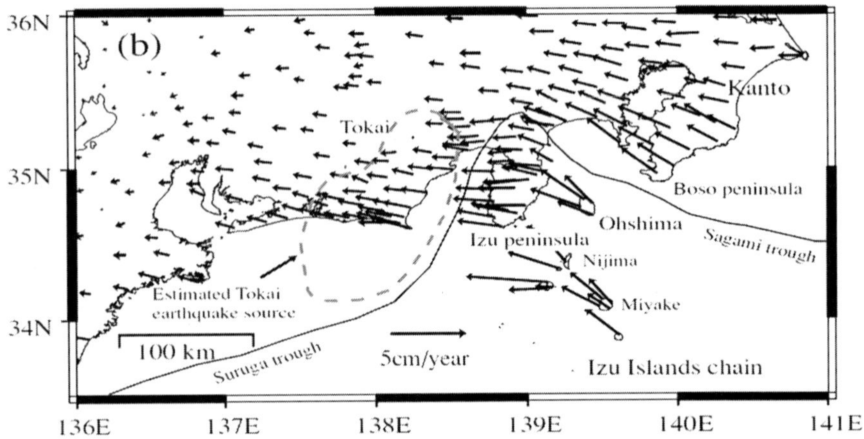

Figure 10. Horizontal deformations in Tokai area, measured by GPS. The Philippine sea plate subducts beneath the continental plate from the Suruga and Sagami troughs. The black arrow represents the observed ground displacement rate in cm/year for the period between 1997 and 1999. Stroke line represents the estimated source area of the expected Tokai earthquake [26].

GPS data show that the Sumatra earthquake of 26 December 2004 (M = 9.0) was generated by a rupture of the Sunda subduction megathrust over a distance of >1,500 kilometers and a width of <150 kilometers. Megathrust slip exceeded 20 meters offshore northern Sumatra, mostly at depths shallower than 30 kilometers. Comparison of the geodetically and seismically inferred slip distribution indicates that approximately 30 % additional fault slip accrued in the 1.5 months following the 500-second-long seismic rupture [27].

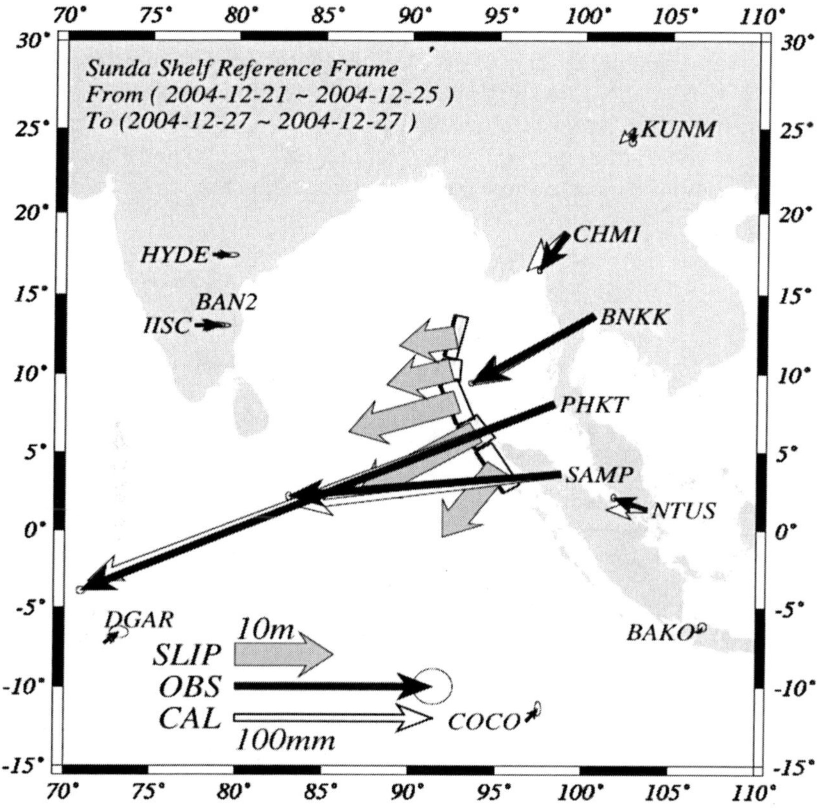

Figure 11. Co-seismic displacements from the Sumatra earthquake (M = 9.0, 26 December 2004, Indonesia). Solid and open arrows show observed by GPS and calculated displacements for the optimal model, respectively. Rectangles are surface projections of assumed fault segments whose upper margins are indicated by thick lines. Thick arrows are the estimated slip on each segment [28].

2.4. Gravity

The detection of an earthquake by a space-based gravitation measurement was reported in 2006 [29]. The Gravity Recovery and Climate Experiment (GRACE) satellites observed a ± 15-microgalileo gravity change induced by the 2004 Sumatra earthquake (Figure 12). Co-seismic deformation produces sudden changes in the gravity field by vertical displacement of Earth's layered density structure and by changing the densities of the crust and mantle. GRACE's sensitivity to the long spatial wavelength of gravity changes resulted in roughly equal contributions of vertical displacement and dilatation effects in the gravity measurements. The GRACE observations provide evidence of crustal dilatation resulting from an undersea earthquake. GRACE is more sensitive (~50% signal retained) to the dilatation effect than to the vertical gravity change (a few %) from the Sumatra EQ co-seismic signals. The negative effect is due to density changes in the crust and the mantle, largely due to expansion of the crust.

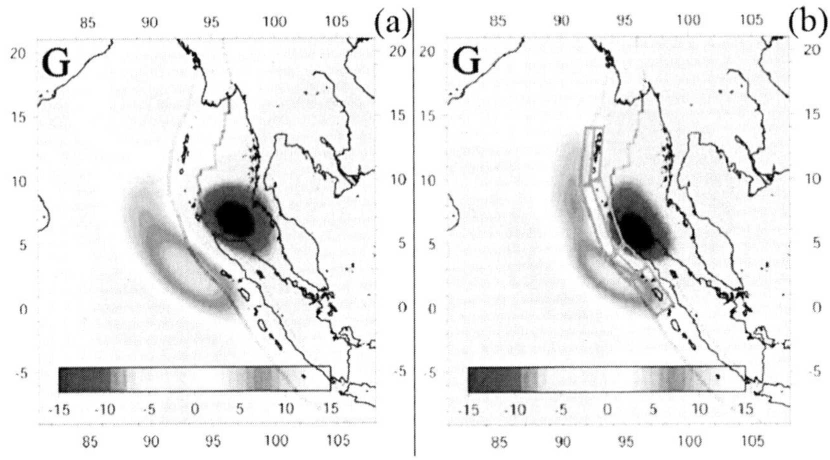

Figure 12. Satellite data showing the gravity changes for the Sumatra earthquake (M = 9.0, 26 December 2004, Indonesia). (a) A composite image from GRACE observed gravity change (±15 μgal). (b) Seismic model predicted gravity change at GRACE resolution [29].

3. THERMAL PHENOMENA

The modern operational space-borne sensors in the TIR spectrum allow monitoring of the Earth's thermal field with a spatial resolution of 0.5–5 km and with a temperature resolution of 0.12–0.5 °C. Surveys are repeated every 12 hours for the polar orbit satellites, and 30 minutes for geostationary satellites. The operational system of polar orbit satellites (two–four satellites in orbit) provides a whole globe survey at least every 6 hours or more frequently. Such sensors may closely monitor seismic prone regions and provide information about the changes in surface temperature associated with an impending earthquake.

Natural phenomena and data availability stimulated the analysis of the long time series of thermal images in relation to earthquake hazard. Historically, the first application of thermal images in earthquake study was carried out in the '80s for Middle Asia [30,31]. Later similar research was carried out in China [32,33], Japan [34], India [35,36], Italy [37], Spain and Turkey [38], USA [39] and other countries. Thermal observations from satellites indicate a significant change of the Earth's surface temperature and near-surface atmosphere layers. Significant thermal anomalies prior to earthquakes related to high seismic areas have been reported for all areas listed above.

3.1. Earth Surface

Satellite observations show the presence of TIR anomalies on the Kamchatka peninsula, Far East Russia, remarkable for its high seismicity, volcanic activity, geothermal fields, geysers, complex relief and bad weather conditions for satellite observations [40]. One example of clear thermal anomaly, related with the 21 Jun 1996 earthquake in Kamchatka is described below. A week before the shock the background distribution of land surface temperature (LST) was observed (Figure 13a). Immediately after the shock 22 Jun 1996 a large scale anomaly was detected in the basin of Kamchatka river (Figure 13b) at the centre of the peninsula. The valley of Kamchatka River is a famous artesian basin with numerous hot springs situated along the 'thermal lines'. Ground observations confirmed satellite data: water temperature and debit of hot springs changed simultaneously with thermal anomaly variation [40].

The destructive Bam earthquake in Iran took place on 26 December 2003. The magnitude of the earthquake was 6.6 and the city of Bam was destroyed. Analyses of the daytime and nighttime datasets for the Bam region showed a distinct anomaly in LST which appeared before the main shock [41]. The temperature increase was about 5–7 °C above the usual temperature of the region. At some places, the temperature was about 6–10 °C higher than the normal temperature of the region in that period of the year. In the nighttime maps, it was seen that on 18 December 2003 there was a complete normal temperature regime in the region (Figure 14). The

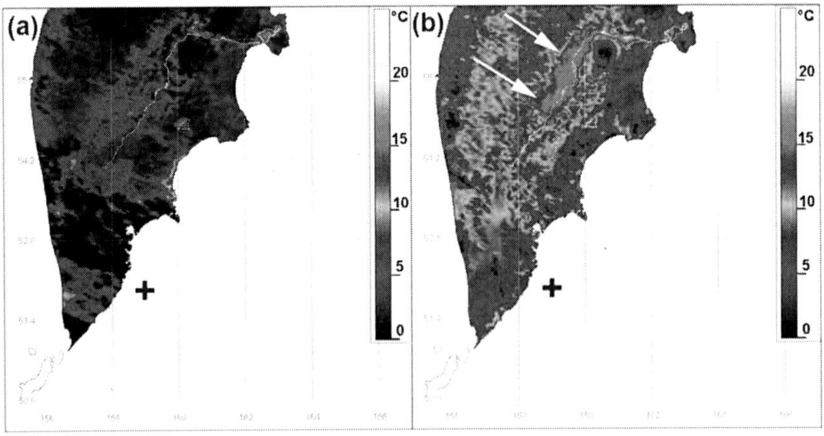

Figure 13. Thermal anomaly related with Kamchatka earthquake (M = 7.0, 21 Jun 1996, Far East, Russia). (a) Background situation, satellite NOAA-14, 14 Jun 1996, cross-earthquake epicenter 21 Jun 1996. (b) Thermal anomaly, satellite NOAA-14, 22 Jun 1996, arrows show thermal anomaly, cross-earthquake epicenter [40].

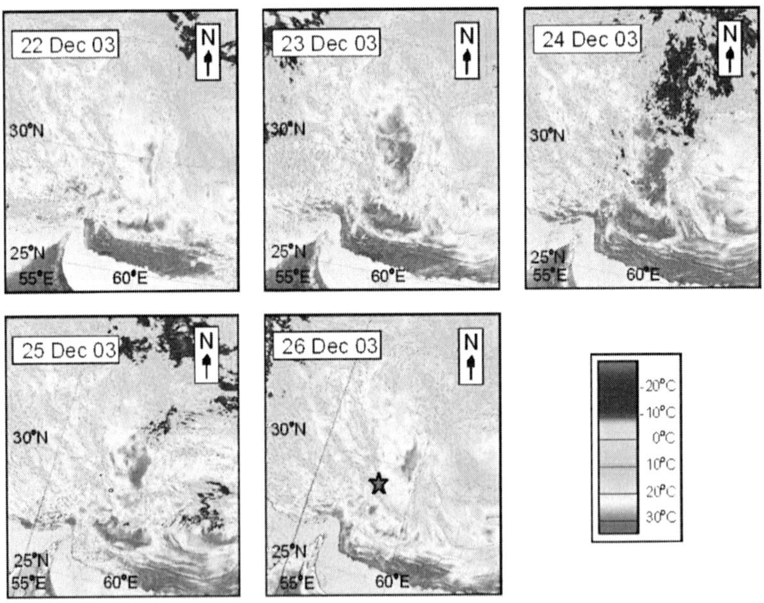

Figure 14. Thermal anomalies associated with Bam earthquake (M = 6.6, 26 December 2003, Iran). NOAA-AVHRR daytime data, star indicate earthquake epicenter [41].

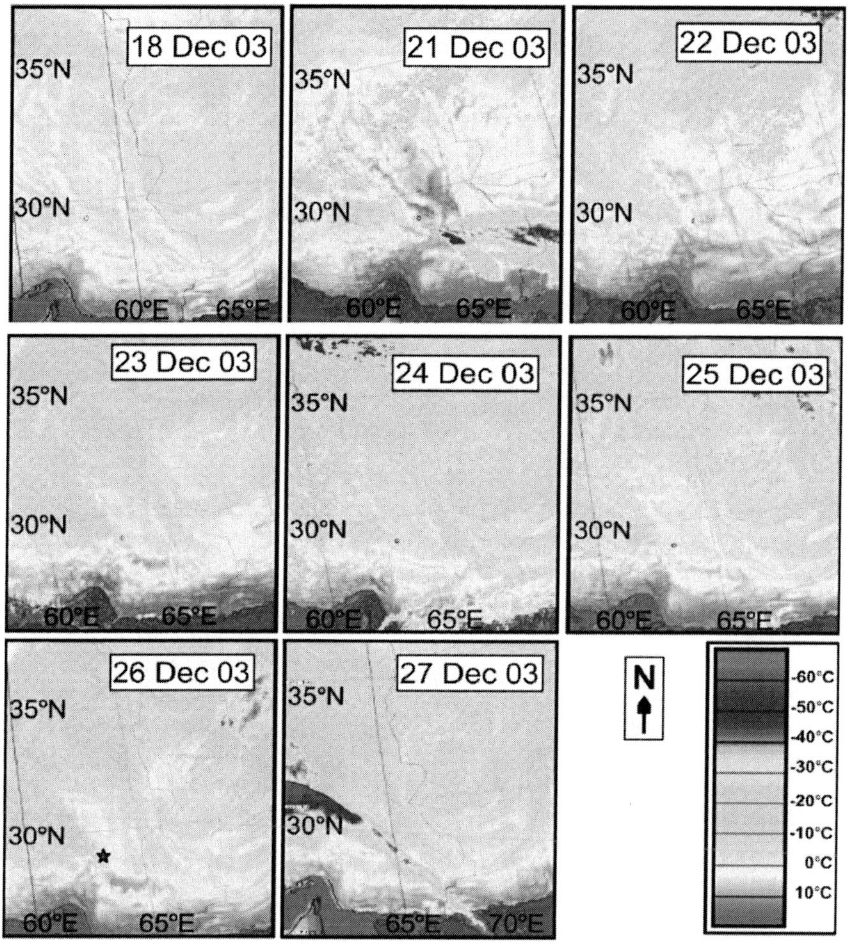

Figure 15. Thermal anomalies associated with the Bam earthquake (M = 6.6, 26 December 2003, Iran). NOAA-AVHRR nighttime data, the star indicates the earthquake epicenter [41].

appearance of an intense thermal anomaly was seen around the earthquake epicenter near Bam on 21 December 2003 (data was not available on 19 and 20 December 2003 for the same time of acquisition). Daytime LST time series maps show that the rise in temperature started on 22 December 2003 (Figure 15). The anomaly remained till 24 December 2003 (just two days before the earthquake). The normal temperature was around 22–25 °C on 21 December 2003. On 24 December 2003, the temperature was around

29–32 °C (about 7–10 °C higher than usual temperature of the region around that period of the year).

Tramutoli and colleagues have developed a robust satellite data analysis technique for space-time thermal anomalies on the Earth's surface recorded by satellites months to weeks before the occurrence of earthquakes [37]. A robust satellite data analysis technique permits identification of space-time thermal infrared anomalies even under very variable observational and natural conditions. A robust estimator of TIR anomalies (RETIRA), ⊗(r,t), where r represents geographic coordinates of the image pixel center, t—is the time of acquisition of the satellite image, has been proposed first by Tramutoli [42]. In this way ⊗(r,t) gives the local excess of the current thermal signal compared with its historical mean value and weighted by its historical variability at the considered location.

A statistically well-founded definition of thermal anomalies is given and proposed as a suitable tool for satellite thermal infrared surveys in seismically active regions. Eight years of Meteosat TIR observations have been analyzed in order to characterize the TIR signal behavior at each specific observation time and location. Space-time TIR signal transients have then been analyzed, both in the presence (validation) and in the absence of (confutation) seismic events, looking for possible space-time relationships. The devastating earthquake which occurred in Izmit, Turkey (M = 7.8, 17 August 1999) has been considered as a test case for validation (Figure 16). Quite intense (Signal/Noise > 3.5) and rare, spatially extensive and time persistent, TIR signal transients were identified, appearing eight days before the Izmit main shock in Greece (August 9th and 12th) and after the shock (August 19th and 20th).

Numerous thermal anomalies related with seismic activity were reported for other resent earthquakes: Colima, M = 7.8, 21 January 2003, Mexico [44]; Bhuj (Gujarat), M = 7.7, 26 January 2001, India [44,45]; Kashmir, M = 7.6, 8 October 2005, [46]; Sichuan, M = 8.0, 12 May 2008, China [47] and many others.

Cases studies of various thermal remote sensing methods applications for earthquakes were reported recently. Pinty et al. [48] have found a significant emergence of surface moisture growth after the Bhuj (Gujarat) earthquake, 26 January 2001, using MISR (Multi-angle Imaging SpectroRadiometer) data. Dey and Singh [49] found significant Surface Latent Heat Flux (SLHF) changes prior to the same earthquake. SLHF anomalies were also reported for the Colima earthquake, M = 7.8, 21 January 2003, Mexico [50] and the Tokachi-Oki earthquake, M = 8.3, 25 September 2003, Japan [51]. Reanalysis of climatic data are also involved in thermal data processing in seismic processes [52].

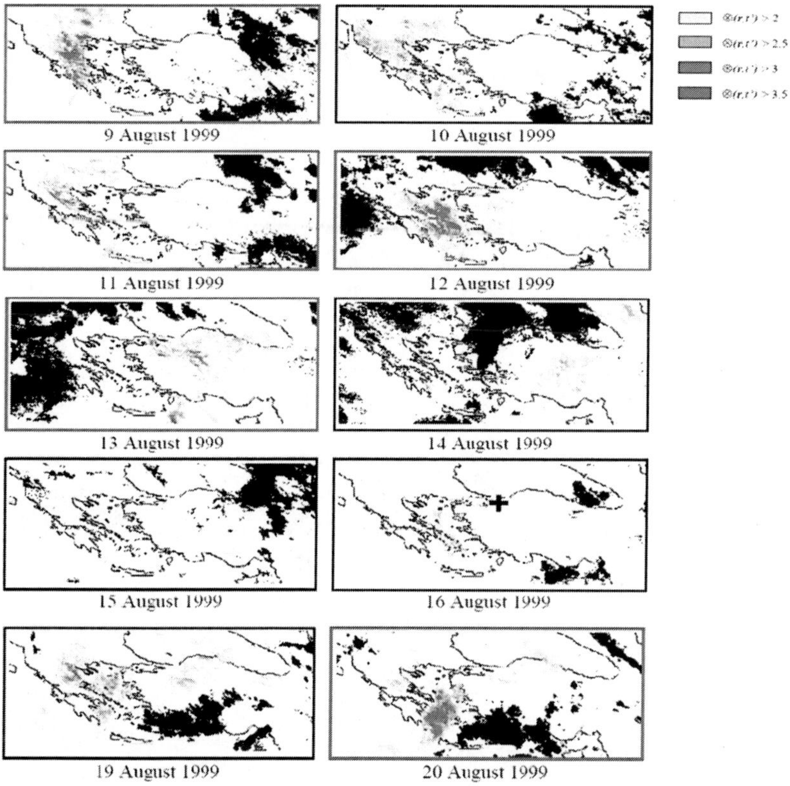

Figure 16. Results of the RETIRA index computation during August 1999 for the Izmit earthquake (M = 7.8, 17 August 1999, Turkey). Cloudy locations (no data) are depicted in black. Pixels with ⊗(r,t) > 3.5 are depicted in red. The black cross, on August 16th, indicates the epicentral area of the Izmit earthquake. Red boxes contour images showing pixels with ⊗(r,t) > 3.5. All scenes refer to Meteosat TIR observations collected at 24:00 GMT) [43].

Satellite thermal survey has a relatively long history of applications in seismology. Most of its advantages and deficiencies are well known. Two main problems limit thermal data utilization: cloud penetration and geological situation. The first issue will be overcome with microwave radiometers in the future. Spatial resolution of microwave devices is too coarse now. The second problem is a fundamental one. There are no strong correlations between seismic activity and thermal anomalies on the Earth's surface. More accurately, we have correlations for some area, for not for others. Surface conditions such as vegetation, precipitations, wind, etc. also strongly influence thermal effects.

3.2. Sea Surface

The sea surface is considered as a more complex object for satellite data applications for seismology. The sea surface temperature depends strongly on weather conditions and sea currents. Due to high thermal inertia all processes of water heating and cooling take place much slower than on the ground.

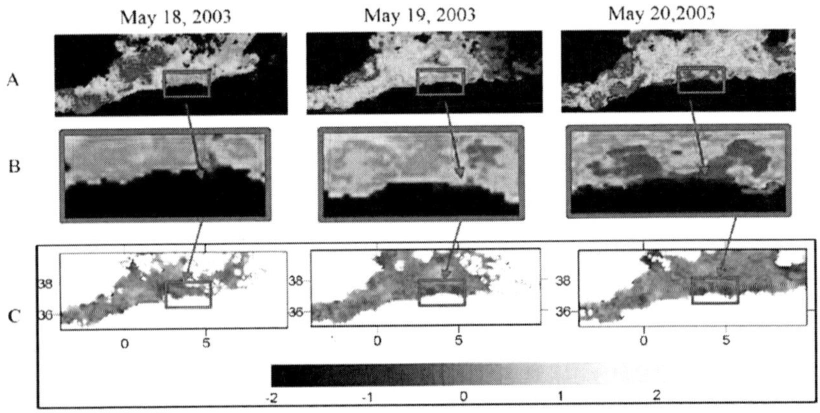

Figure 17. Sea surface temperature variation for Boumerdes earthquake (M = 6.8, 21 May 2003, Algeria). MODIS Terra LST for May 18, 19 and 20 in near zone to the epicenter. (a) Map of 4.5 km nighttime girded SST data, with red rectangle marking the upwelling area. Land is masked with black color. (b) Zoom in the epicentral area with cooling the surface temperature (blue color). (c) SST residual computed for the same days by the method of median polish [44].

Sea surface temperatures (SST) were analyzed for the Boumerdes earthquake (northern Algeria, 21 May 2003, M = 6.8 [44]). Statistical analysis was preformed to estimate the SST variability during the period of 90 days around 21 May 2003. Night images of Terra/MODIS for May 18, 19 and 20 were shown as the land was masked with black (Figure 17). The decrease of SST with amplitude –2 °C was observed on locations close to the epicenter and this temperature decrease pattern could not be seen before or after the main shock. The statistical analysis confirms that a few days before the main shock there were local up-wellings of cold water with an average SST –2 °C in comparison with the background mean temperature. Some other examples of SST anomalies, mostly negative, related with earthquake were reported by Ouzounov and Freund [53].

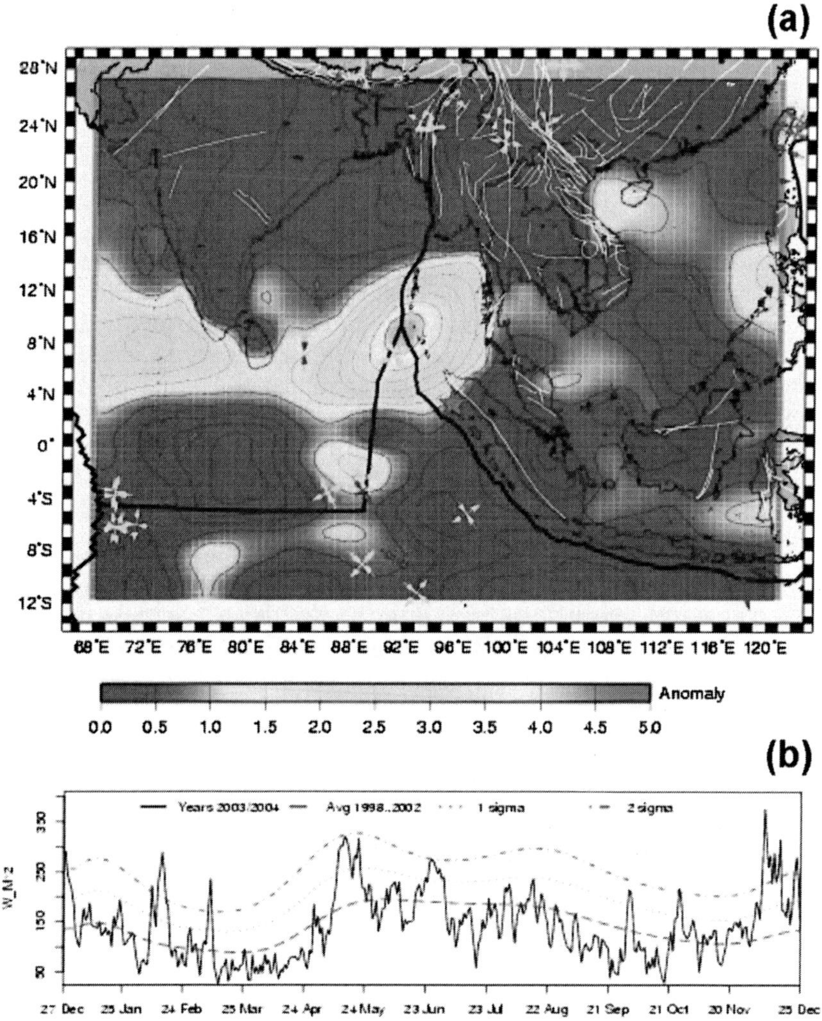

Figure 18. Surface latent heat flux anomalies for the Sumatra earthquake (M = 9.0, 26 December 2004, Indonesia). (a) SLHF anomaly on 7 December 2004. (b) Time series of wavelet analysis of SLHF for the period from 27 December 2003 to 25 December 2004 [55].

Singh et al. [54] found changes in water color and Earth's surface related to Bhuj (Gujarat) earthquake of January 26 using IRS satellite data. Numerous sea surface parameters variations were discussed in relation with Sumatra earthquake, 2004, from SLHF (Figure 18) till chlorophyll-a concentrations [55].

3.3. Atmosphere

Thermal anomalies before strong earthquakes are observed at different levels, starting from the ground surface up to the top of clouds altitude. All kinds of anomalies were registered repeatedly by many researchers before different earthquakes by remote sensing satellite techniques on the earth and sea surface. The most promising in the present moment is the Outgoing Longwave Radiation (OLR) anomaly measured at the top of clouds level [56]. Its advantage is that it is measured within the infrared transparency window of 8–12 microns and does not separate clouds. OLR is currently mapped by the AIRS (Atmospheric Infrared Sounder) launched into orbit in 2002. AIRS is one of six instruments on board the Aqua satellite, part of the NASA Earth Observing System. One of the best examples of OLR anomalies is related with the Sumatra earthquake in 2004 (Figure 19).

It is difficult to expect that something like a ground surface temperature increase by 3–5 degrees or infrared emission emitted by rocks due to deformation can provide at an altitude of 12 km the energy flux within the range 4–80 W/m2 which was measured experimentally. Such an amount of energy could only be provided by some process taking place directly in the atmosphere.

Dey et al. [57] found changes in the total water vapour column after the Bhuj (Gujarat) earthquake. Water content was retrieved by SSM/I microwave radiometer on the Tropical Rainfall Measuring Mission (TRMM) satellite. Okada et al. [58] found changes in atmospheric aerosol parameters after the Bhuj (Gujarat) earthquake. Aerosols above the sea surface were revealed on the base SeaWiFS satellite data. A few examples of ozone concentrations changes measured by TOMS related with earthquakes have been reported by Tronin [59].

Also we can mention about numerous attempts to use cloud detection for earthquake research [60]. Cloud anomalies were detected (Figure 15) on satellite images before (21 December 2003) and after (27 December 2003) the Bam earthquake, Iran [61] and for some Japanese earthquakes [62].

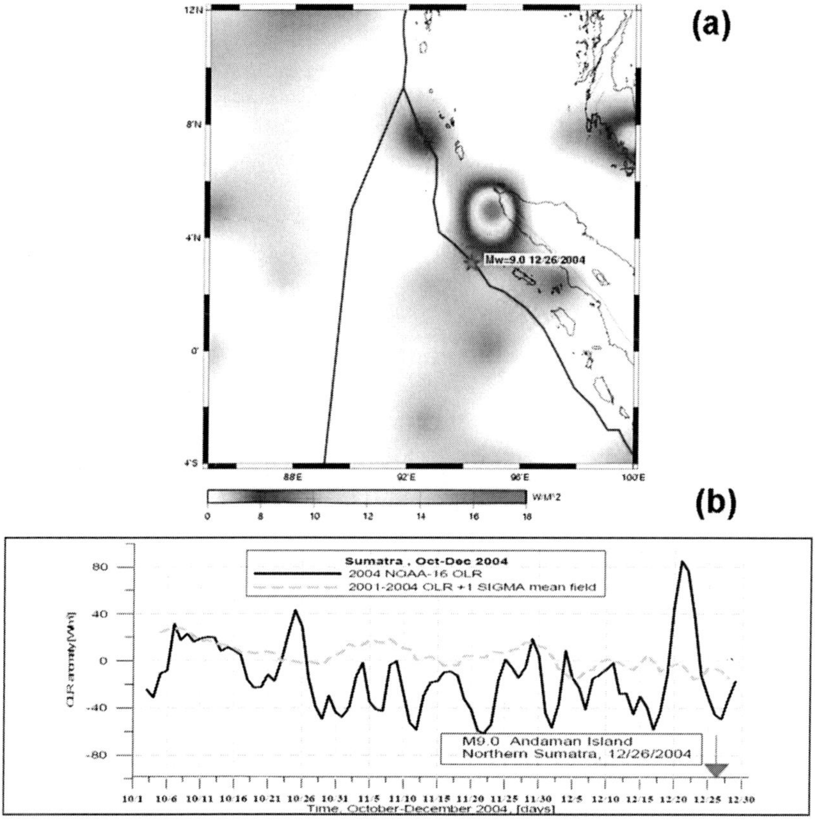

Figure 19. Outgoing longwave radiation anomaly for the Sumatra earthquake (M = 9.0, 26 December 2004, Indonesia). (a) Map of OLR monthly variations for November 2004, month prior to the shock. (b) Time-series of daily OLR anomaly for 1 October 2004–31 December 2004 over the epicenter [56].

4. ELECTROMAGNETICS

Electromagnetic observations do not belong to remote sensing. Nevertheless the role of these methods in earthquake research is very important. One can find a comprehensive review on seismo-electromagnetics presented by Uyeda, Nagao and Kamogawa [63]. The DEMETER mission [64] successfully executes a science program on seismo-ionoshperic research [65,66].

Total electron content (TEC) is one the main parameters studied in relation with earthquakes [6]. GPS observations at the moment are the main tool for TEC measurements. The density of GPS stations increases every year, which allows for restoring the 3D ionosphere structure. An excellent example of TEC variation (Figure 20) was represented for the Chi-Chi earthquake [67]. Modern TEC applications can be found in the publications of Liu [68,69,70]. On the other hand the constellation of satellites with scatterometers like QuikSCAT generates TEC images above sea surface as side effects.

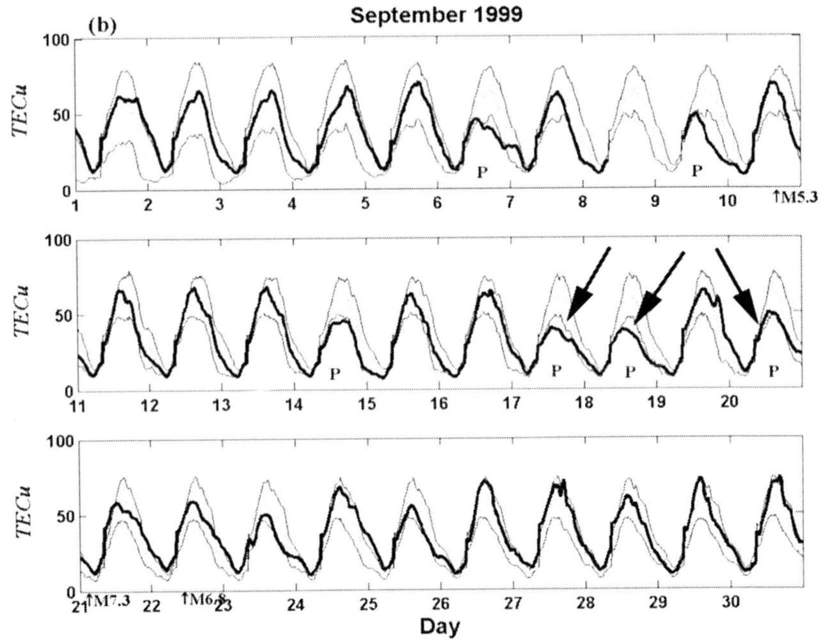

Figure 20. The variations of TEC observed in September 1999 for the Chi-Chi earthquake (M = 7.6, 20 September 1999, Taiwan). 1 TECu = 1016 el/m2. Arrows show TEC anomalies [67].

5. DISCUSSION

A wide spectrum of satellite remote sensing methods are applied in seismology nowadays. The value of these methods for earthquake research is varied. Optical methods have limited applications, mostly for rapid assessment of damages in an epicentral zone. Other applications such as alignment analysis and cloud form analysis related with earthquakes do not have an adequate scientific basis for seismological application. Vigorous extension of InSAR methods applications in seismology is observed now. Modern radar systems in conjunction with GPS/GLONASS will provide whole seismic cycle monitoring. Broad application of InSAR methods is limited by the high data cost and complex data analysis. Thermal satellite data applications are developing in two directions at the moment: thermal anomalies in seismic fault research and emitted longwave radiation measurements in seismic zones. Thermal anomalies research in seismic faults is developing in the direction of seismic activity monitoring and close integration with ground observations. Emitted longwave radiation observations demonstrate promising results, but data accumulation is required. The nature of ongoing longwave radiation anomalies remains unclear.

Some common remarks on satellite data application in seismology can be made: (1) The level of automatic data processing is insufficient. There is still too much manual labour and author arbitrariness in data processing—this concerns both exotic earthquake cloud analysis and high technique radar methods. Some results are irreproducible. (2) There is a weak physical and geological basis for many of the proposed methods. The nature and driving forces of some phenomena need clarification and connection with current understanding of physics and geology.

6. CONCLUSIONS

This review of modern remote sensing techniques in seismology demonstrates the following: (1) remote sensing methods are being broadly used for earthquake research; (2) a wide spectra of remote sensing methods are applied—from optical sensors to radar systems; (3) the list of parameters studied by remote sensing are: surface deformation (both vertical and horizontal), surface temperature, various heat fluxes on the Earth's and top clouds surfaces and some others; (4) future development of remote sensing application for earthquakes related with new directions: L-band radar systems, high-resolution microwave radiometers, gas analyzers; (5) we will probably again approach an epoch of "belief" in earthquake prediction, where remote sensing can play a key role due to its global scope, calibration, and automatic data processing.

The described processes in the ionosphere, atmosphere, hydrosphere and lithosphere associated with earthquakes represent the fundamental science issue of lithosphere-atmosphere-ionosphere coupling. The solution of this problem is quite far away. We can mention specifically the problems of the nature of thermal anomalies, the nature of emitted longwave radiation anomalies, ionosphere-lithosphere coupling and so on. All these issues interface with the problem of understanding the nature of earthquakes.

REFERENCES

1. Showalter, P.S. Remote sensing's use in disaster research: a review. Disaster Prevent. Manage. 2001, 10, 21–29.
2. Paylor, E.D., II; Evans, D.L.; Tralli, D.M. Remote sensing and geospatial information for natural hazards characterization. ISPRS J. Photogramm. Remote Sens. 2005, 59, 181–253.
3. IGOS. International Global Observation Strategy Geohazards Theme Report; ESA/UNESCO: European Space Agency, Paris, France, 2004; p. 54.
4. Joyce, K.E.; Belliss, S.E.; Samsonov, S.V.; McNeill, S.J.; Glassey, P.J. A review of the status of satellite remote sensing and image processing techniques for mapping natural hazards and disasters. Prog. Phys. Geog. 2009, 33, 183–207.
5. Tronin, A.A. Remote sensing and earthquakes: a review. Phys. Chem. Earth 2006, 31, 138–142.
6. Hayakawa, M. Atmospheric and Ionospheric Electromagnetic Phenomena Associated with Earthquakes; Terra Sci. Pub. Co. Ltd.: Tokyo, Japan, 1999; p. 996.
7. Seismo Electromagnetics: Lithosphere-Atmosphere-Ionosphere Coupling; Hayakawa, M., Molchanov, O.A., Eds.; Terra Science Publishing Co. Ltd.: Tokyo, Japan, 2002; p. 477.
8. Clark, M.M. Finding active faults using aerial photographs. Earthqua. Inf. Bull. 1978, 10, 169–173.
9. Trifonov, V.G. Active faults in Eurasia: general remarks. Tectonophysics 2004, 380, 123–130.
10. Trifonov, V.G.; Makarov, V.I.; Kojurin, A.I.; Skobelev, S.F.; Schulz, S.S., Jr. Aerospace Research of Seismic Zones; Nauka: Moscow, USSR, 1988; p. 133.

11. Saraf, A.K. IRS-1C-PAN Depicts Chamoli earthquake induced landslides in Garhwal Himalayas, India. Int. J. Remote Sens. 2000, 21, 2345–2352.

12. Van Puymbroeck, N.; Michel, R.; Binet, R.; Avouac, J.P.; Taboury, J. Measuring earthquakes from optical satellite images. Appl. Opt. Inf. Process. 2000, 39, 1–14.

13. Michel, R.; Avouac, J.P. Deformation due to the 17 August Izmit, Turkey, earthquake measured from SPOT images. J. Geophys. Res. 2002.

14. Dominguez, S.; Avouac, J.P.; Michel, R. Horizontal coseismic deformation of the 1999 Chi-Chi earthquake measured from SPOT satellite images: Implications for the seismic cycle along the western foothills of Central Taiwan. J. Geophys. Res. 2003.

15. Arellano-Baeza, A.A.; Zverev, A.T.; Malinnikov, V.A. Study of changes in the lineament structure, caused by earthquakes in South America by applying the lineament analysis to the Aster (Terra) satellite data. Adv. Space Res. 2006, 37, 690–697.

16. Massonnet, D.; Rossi, M.; Carmona, C.; Adragna, F.; Peltzer, G.; Feigl, K.; Rabaute, T. The displacement field of the Landers earthquake mapped by radar interferometry. Nature 1993, 364, 138–142.

17. Peltzer, G.; Hudnut, K.; Feigl, K. Analysis of coseismic surface displacement gradients using radar interferometry: New insights into the Landers earthquake. J. Geophys. Res. 1994, 99, 21971–21981.

18. Tobita, M.; Fujiwara, S.; Ozawa, S.; Rosen, P.A.; Fielding, E.J.; Werner, C.L.; Murakami, M.; Nakagawa, H.; Nitta, K.; Murakami, M. Deformation of the 1995 North Sakhalin earthquake detected by JERS-1/SAR interferometry. Earth Planets Space 1998, 50, 313–325.

19. Schmidt, D.A.; Bürgmann, R. InSAR constraints on the source parameters of the 2001 Bhuj earthquake. Geophys. Res. Lett. 2006, 33, L02315.

20. Stramondo, S.; Moro, M.; Tolomei, C.; Cinti, F.R.; Doumaz, F. InSAR surface displacement field and fault modelling for the 2003 Bam earthquake (southeastern Iran). J. Geodynamics 2005, 40, 347–353.

21. Chini, M.; Bignami, C.; Stramondo, S.; Pierdicca, N. Uplift and subsidence due to the 26 December 2004 Indonesian earthquake detected by SAR data. Int. J. Remote Sens. 2008, 29, 3891–3910.

22. Kuzuoka, S.; Mizuno, T. Land Deformation Monitoring Using PSInSAR Technique. In Proceedings of International Symposium on Monitoring, Prediction and Mitigation of Disasters by Satellite Remote Sensing, Awaji, Hyogo, Japan, January 2004; pp. 176–181.

23. Tsai, Y.B.; Liu, J.Y.; Ma, K.F.; Yen, H.Y.; Chen, K.S.; Chen, Y.I.; Lee, C.P. Precursory phenomena associated with the 1999 Chi-Chi earthquake in Taiwan as identified under iSTEP Program. Phys. Chem. Earth 2006, 31, 365–377.

24. de Michele, M.; Raucoulesa, D.; Salichon, J.; Lemoine, A.; Aochia, H. Using INSAR for seismotectonic observations over the Mw 6.3 Parkfield earthquake (28/09/2004), California. In Proceedings of Commission IV ISPRS Congress, Beijing, China, July 2008; Available online: http://www.isprs.org/congresses/beijing2008/proceedings/4_pdf/265.pdf (accessed on 15 November 2009).

25. Fielding, E.J.; Lundgren, P.R.; Bürgmann, R.; Funning, G.J. Shallow fault-zone dilatancy recovery after the 2003 Bam, Iran earthquake. Nature 2009, 458, 64–68.

26. Ozawa, S.; Murakami, M.; Kaidzu, M.; Hatanaka, Y. Transient crustal deformation in Tokai region, centeral Japan, until May 2004. Earth Planets Space 2005, 57, 909–915.

27. Subarya, C.; Chlieh, M.; Prawirodirdjo, L.; Avouac, J.-P.; Bock, Y.; Sieh, K.; Meltzner, A.J.; Natawidjaja, D.H.; McCaffrey, R. Plate-boundary deformation associated with the great Sumatra-Andaman earthquake. Nature 2006, 440, 46–51.

28. Hashimoto, M.; Choosakul, N.; Hashizume, M.; Takemoto, S.; Takiguchi, H.; Fukuda, Y.; Fujimori, K. Crustal deformations associated with the great Sumatra-Andaman earthquake deduced from continuous GPS observation. Earth Planets Space 2006, 58, 127–139.

29. Han, S.-C.; Shum, C.K.; Bevis, M.; Ji, C.; Kuo, C.-Y. Crustal Dilatation Observed by GRACE after the 2004 Sumatra-Andaman Earthquake. Science 2006, 313, 658–662.

30. Gorny, V.I.; Salman, A.G.; Tronin, A.A.; Shilin, B.V. The earth's outgoing IR radiation as an indicator of seismic activity. Proc. Acad. Sci. USSR 1988, 301, 67–69.

31. Tronin, A.A. Satellite thermal survey—a new tool for the studies of seismoactive regions. Int. J. Remote Sens. 1996, 17, 1439–1455.

32. Quang, Z.; Xu, X.; Dian, C. Thermal infrared anomaly precursors of impending earthquakes. Chin. Sci. Bull. 1991, 36, 319–323.

33. Qiang, Z.; Du, L.-T. Earth degassing, forest fire and seismic activities. Earth Science Frontiers 2001, 8, 235–245.

34. Tronin, A.A.; Hayakawa, M.; Molchanov, O.A. Thermal IR satellite data application for earthquake research in Japan and China. J. Geodynamics 2002, 33, 519–534.

35. Saraf, A.K.; Choudhury, S. Earthquakes and thermal anomalies. Geospatial Today 2003, 2, 18–20. [Google Scholar]

36. Singh, R.P.; Ouzounov, D. Earth processes in wake of Gujarat earthquake reviewed from space. EOS Trans. Amer. Geophys. Union 2003, 84, 244.

37. Tramutoli, V.; Bello, G.D.; Pergola, N.; Piscitelli, S. Robust satellite techniques for remote sensing of seismically active areas. Ann. Geofis. 2001, 44, 295–312.

38. CORDIS RTD-PROJECTS. Regular Update of Seismic Hazard Maps through Thermal Space Observations; European Communities: Madrid, Spain, Project Reference: ENV4980741; 2000.

39. Ouzounov, D.; Freund, F. Mid-infrared emission prior to strong earthquakes analyzed by remote sensing data. Adv. Space Res. 2003, 33, 268–273.

40. Tronin, A.A.; Biagi, P.F.; Molchanov, O.A.; Khatkevich, Y.M.; Gordeeev, E.I. Temperature variations related to earthquakes from simultaneous observation at the ground stations and by satellites in Kamchatka area. Phys. Chem. Earth 2004, 29, 501–506.

41. Saraf, A.K.; Rawat, V.; Banerjee, P.; Choudhury, S.; Panda, S.K.; Dasgupta, S.; Das, J.D. Satellite detection of earthquake thermal precursors in Iran. Nat. Hazard 2008, 47, 119–135.

42. Tramutoli, V. Robust AVHRR Techniques (RAT) for environmental monitoring: theory and applications. In Earth Surface Remote Sensing II; Cecchi, G., Zilioli, E, Eds.; SPIE: Barcelona, Spain, 1998; pp. 101–113. [Google Scholar]

43. Tramutoli, V.; Cuomo, V.; Filizzola, C.; Pergola, N.; Pietrapertosa, C. Assessing the potential of thermal infrared satellite surveys for monitoring seismically active areas. The case of Kocaeli (İzmit) earthquake, August 17, 1999. Remote Sens. Environ. 2005, 96, 409–426.

44. Ouzounov, D.; Bryant, N.; Logan, T.; Pulinets, S.; Taylor, P. Satellite thermal IR phenomena associated with some of the major earthquakes in 1999–2003. Phys. Chem. Earth 2006, 31, 154–163.

45. Saraf, A.K.; Choudhury, S. NOAA-AVHRR detects thermal anomaly associated with 26 January, 2001 Bhuj Earthquake, Gujarat, India. Int. J. Remote Sens. 2005, 26, 1065–1073.

46. Panda, S.K.; Choudhury, S.; Saraf, A.K.; Das, J.D. MODIS land surface temperature data detects thermal anomaly preceding 08 October 2005 Kashmir earthquake. Int. J. Remote Sens. 2007, 28, 4587–4596.

47. Wei, L.; Guo, J.; Liu, J.; Lu, Z.; Li, H.; Cai, H. Satellite thermal infrared earthquake precursor to the Wenchuan Ms 8.0 earthquake in Sichuan, China, and its analysis on geodynamics. Acta Geol. Sinica-Engl. Ed. 2009, 83, 767–775.

48. Pinty, B.; Gobron, N.; Verstraete, M.M.; Mélin, F.; Widlowski, J.-L.; Govaerts, Y.; Diner, D.J.; Fielding, E.; Nelson, D.L.; Madariaga, R.; Tuttle, M.P. Observing earthquake-related dewatering using MISR/Terra satellite Data. EOS Trans. Amer. Geophys. Union 2003, 84, 37–48.

49. Dey, S.; Singh, R.P. Surface latent heat flux as an earthquake precursor. Nat. Hazards Earth Syst. 2003, 3, 749–755.

50. Pulinets, S.A.; Ouzounov, D.; Ciraolo, L.; Singh, R.; Cervone, G.; Leyva, A.; Dunajecka, M.; Karelin, A.V.; Boyarchuk, K.A.; Kotsarenko, A. Thermal, atmospheric and ionospheric anomalies around the time of the Colima M7.8 earthquake of 21 January 2003. Ann. Geophys. 2006, 24, 835–849.

51. Cervone, G.; Maekawa, S.; Singh, R.; Hayakawa, M.; Kafatos, M.; Shavets, A. Surface latent heat flux and nighttime LF anomalies prior to the Mw = 8.3 Tokachi-Oki earthquake. Nat. Hazards Earth Syst. Sci. 2006, 6, 109–114.

52. Ma, W.; Ma, W.; Zhao, H.; Li, H. Temperature changing process of the Hokkaido (Japan) earthquake on 25 September 2003. Nat. Hazards Earth Syst. Sci. 2008, 8, 985–989.

53. Ouzounov, D.; Freund, F. Mid-infrared emission prior to strong earthquakes analyzed by remote sensing data. Adv. Space Res. 2004, 33, 268–273.

54. Singh, R.P.; Bhoi, S.; Sahoo, A.K. Changes observed on land and ocean after Gujarat earthquake of January 26, 2001 using IRS data. Int. J. Remote Sens. 2002, 23, 3123–3128.

55. Singh, R.P.; Cervone, G.; Kafatos, M.; Prasad, A.K.; Sahoo, A.K.; Sun, D.; Tang, D.L.; Yang, R. Multi-sensor studies of the Sumatra earthquake and tsunami of 26 December 2004. Int. J. Remote Sens. 2007, 28, 2885–2896.

56. Ouzounov, D.; Liu, D.; Kang, C.; Cervone, G.; Kafatos, M.; Taylor, P. Outgoing long wave radiation variability from IR satellite data prior to major earthquakes. Tectonophysics 2007, 431, 211–220.

57. Dey, S.; Sarkar, S.; Singh, R.P. Anomalous changes in column water vapor after Gujarat earthquake. Adv. Space Res. 2004, 33, 274–278.

58. Okada, Y.; Mukai, S.; Singh, R.P. Changes in atmospheric aerosol parameters after Gujarat earthquake of January 26, 2001. Adv. Space Res. 2004, 33, 254–258.

59. Tronin, A.A. Atmosphere-litosphere coupling. Thermal anomalies on the earth surface in seismic processes. In Seismo Electromagnetics: Lithosphere-Atmosphere-Ionosphere Coupling; Hayakawa, M., Molchanov, O.A., Eds.; TERRAPUB: Tokyo, Japan, 2002; pp. 173–176.

60. Morozova, L.I. Features of atmo-lithmospheric relationships during periods of strong Asian earthquakes. Izvestiya, Phys. Solid Earth 1996, 5, 63–68.

61. Guo, G.; Wang, B. Cloud anomaly before Iran earthquake. Int. J. Remote Sens. 2008, 29, 1921–1928.

62. Gup, G.; Xie, G. Earthquake cloud over Japan detected by satellite. Int. J. Remote Sens. 2007, 28, 5375–5376.

63. Uyeda, S.; Nagao, T.; Kamogawa, M. Short-term earthquake prediction: current status of seismo-electromagnetics. Tectonophysics 2009, 470, 205–213.

64. Parrot, M. The micro-satelitte DEMETER. J. Geodynamics 2002, 33, 535–541.

65. Molchanov, O.; Rozhnoi, A.; Solovieva, M.; Akentieva, O.; Berthelier, J.J.; Parrot, M.; Lefeuvre, F.; Biagi, P.F.; Castellana, L.; Hayakawa, M. Global diagnostics of the ionospheric perturbations related to the seismic activity using the VLF radio signals collected on the DEMETER satellite. Nat. Hazards Earth Syst. Sci. 2006, 6, 745–753.

66. Parrot, M.; Berthelier, J.J.; Lebreton, J.P.; Sauvaud, J.A.; Santolík, O.; Blecki, J. Examples of unusual ionospheric observations made by the DEMETER satellite over seismic regions. Phys. Chem. Earth 2006, 31, 486–495.

67. Liu, J.; Chuo, Y.; Shan, S.; Tsai, Y.; Chen, Y.; Pulinets, S.; Yu, S. Pre-earthquake ionospheric anomalies registered by continuous GPS TEC measurements. Ann. Geophys. 2004, 22, 1585–1593.

68. Liu, J.Y.; Tsai, Y.B.; Chen, S.W.; Lee, C.P.; Chen, Y.C.; Yen, H.Y.; Chang, W.Y.; Liu, C. Giant ionospheric disturbances excited by the M9.3 Sumatra earthquake of 26 December 2004. Geophys. Res. Lett. 2006, 33, L02103.

69. Liu, J.-Y.; Tsai, Y.-B.; Ma, K.-F.; Chen, Y.-I.; Tsai, H.-F.; Lin, C.-H.; Kamogawa, M.; Lee, C.-P. Ionospheric GPS total electron content (TEC) disturbances triggered by the 26 December 2004 Indian Ocean tsunami. J. Geophys. Res. 2006, 111, A05303.

70. Liu, J.Y.; Chen, Y.I.; Chen, C.H.; Liu, C.Y.; Chen, C.Y.; Nishihashi, M.; Li, J.Z.; Xia, Y.Q.; Oyama, K.I.; Hattori, K.; Lin, C.H. Seismoionospheric GPS total electron content anomalies observed before the 12 May 2008 Mw7.9 Wenchuan earthquake. J. Geophys. Res. 2009, 114, A04320

5

CHAPTER

THE USE OF SOURCE SCALING RELATIONSHIPS IN THE SIMULATION OF A SEISMIC SCENARIO IN MEXICO

Alejandro Gaytán[1], Carlos I. Huerta Lopez[2], Jorge Aguirre Gonzales[3] and Miguel A. Jaimes[3]

[1] *Universidad de Guadalajara, Departamento de Ciencias Computacionales, Centro Universitario de Ciencias Exactas e Ingeniería, Mexico*

[2] *University of Puerto Rico at Mayaguez Civil Engineering and Surveying Department, Puerto Rico Strong Motion Program Home Institution: Research Center and Higher Education at Ensenada, CICESE Seismology Department, Earth Sciences Division, Puerto Rico*

[3] *Instituto de Ingeniería, Universidad Nacional Autónoma de México, Circuito Interior, Ciudad Universitaria, Mexico*

1. INTRODUCTION

The general knowledge of seismicity of the Mexican Pacific Coast has been described in terms of its relationship within the regional context with the Middle America Trench, the convergence rate observed and predicted by plate tectonics processes, the use of seismic gap theory for forecast earthquakes on the region (Reyes [1]; Eissler and McNally [2]; Singh [3]; as well as Heaton and Kanamori [4]; among others) and prediction of strong ground motions using empirical relationship and seismic source scaling procedures (Somerville [5] and Somerville [6]).

The work here presented, mainly deals with the Colima-Jalisco region (ruptures areas of 1932, and 1973 earthquakes) as well as the remained gap of the Tecoman (2003) earthquake. The conduction of studies in this region to simulate seismic scenarios acquires special significance because of the hazard of an eventual large magnitude earthquake. For the large earthquakes recording in this region last and present century the rupture areas of the 1932, 1973 earthquakes and the small gap remainder from Tecoman earthquake the seismic convergence period have been exceeded or is in its limit.

Let us look into this with detail. Firstly, for the earthquakes of 3 and 18 June 1932, Nishenko and Singh [7] found that the average displacement on the fault of 1932 earthquakes as 155 cm. Pardo and Suarez [8] found that subduction rate of Rivera plate in this region is estimated to be from 2 to 5 cm/yr, take the small estimation of convergence rate of 2 cm/yr yields a recurrence period of about 77 years assuming that the convergence is entirely taken up by the seismic slip (Nishenko and Singh [7]). However considering a rate larger than 2 cm/yr the recurrence period could be much lower. The 79 years elapsed since 1932 indicate that the recurrence period has expired. Although the rupture area of Manzanillo 1995 Mw 8.0 earthquake invades the rupture area of 1932 earthquakes, but represent less than 15 % of rupture area. Secondly, for the January 30 1973 earthquake (Mw 7.3) Reyes [1] estimate an average slip of 144 cm. For this region Minster [9] estimate a convergence rate of 5.6 cm/yr, this suggest a recurrence time period of about 25 years Reyes [1]. This recurrence period consider the various uncertainties and is in very good agreement with the interval between this earthquake and the preceding 1941 earthquake (32 years). The 38 years trascurred since 1973 indicate that the recurrence period have expired Reyes [1].

Thirdly, is the existence of a seismic gap in the region located between the rupture areas of the 1973 and the Tecoman earthquakes. In figure 1 Quintanar [12] shows that the aftershocks location of the Tecoman earthquake lies north of El Gordo graben and the aftershock area encompass part of the rupture area of the 1932 and 1995 earthquakes. The area between the limits of the rupture areas of the 1995 and the1973 earthquakes is what has been called the Colima seismic gap. The northwest area of this gap ruptured with the Tecoman earthquake in 2003. The other half of the gap, roughly to the southeast, remains quiet Quintanar [12].

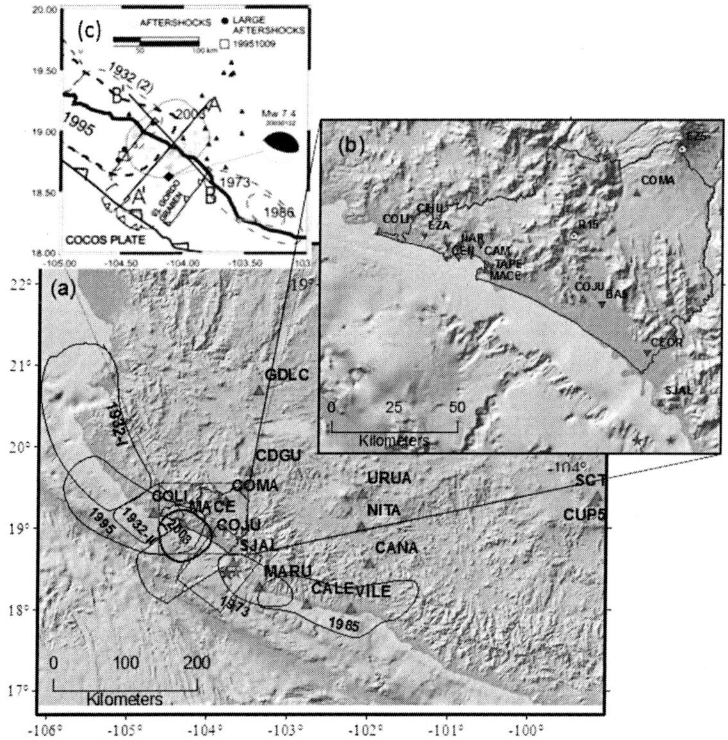

Figure 1. Geographic regional map. (a) Contours show historic earthquakes in region (modified from Singh [10], local and regional stations where the 13 August 2006 earthquake (small star) were recorded: red triangles, big star epicentral location of simulated earthquake. (b) Local and regional stations within Colima state where earthquake was recorded; blue triangles, IINGEN array and pentagons, RESCO array. (c) From Quintanar [11], map showing the best aftershock locations (empty circles). Large full square, shows the epicenter location of the Tecoman earthquake. Its fault plane solution is also shown. Large empty square shows the epicenter location of the Colima–Jalisco earthquake of 1995 (Mw = 8.0). Profiles AA' and BB' are cross sections onto which aftershocks were projected. Triangles show the locations of seismic stations.

The later show the existence of 3 different zones where is necessary to make a simulation of strong ground motions to estimate the response spectra. From these 3 zones the more important (by magnitude and area) is the region broken by earthquakes of 1932; however, the possibility of conduct a simulation in this area is limited by the poor instrumentation in the past and therefore the absence of data the region.

In the case of small gap remainder from Tecoman and the rupture area of 1973 earthquakes, the Mw 5.3 earthquake of August 13, 2006 (element event) offered the opportunity to generate a seismic scenario for the studied area that constitutes a potential seismic source in this region in the near future. We apply: (i) the empirical Green's function method (EGFM) proposed by Irikura [13] to estimate the peak ground acceleration (PGA) and the acceleration time histories using the Somerville [6] relationship allowing us scaling from moderate to major earthquake. Traditionally the EGFM is well known technique commonly used to simulate an already occurred earthquake. The scientific community thinks that would be of great benefit if this technique can be applied to forecast an expected earthquake in Mexico. Under this premise, an immediate doubt arise because the absence of observed records (event we intent simulate) to compare with synthetics. Conduct a study to simulate strong ground motions under this situation makes difficult to constrain and validate the results, but at the same time is one of the more important challenge simulating a future earthquake. In order to overcome the absence of observed records two methodologies were applied: (i) The first one was Somerville [6] relationship together with the EGFM, and (ii) the second one was ground motions predictive equations (GMPE) relationships to make accurate estimation of strong motions. Aguirre and Irikura [14] used acceleration records of Mexican subduction earthquakes to validate the Somerville [5, 6] relationship for subduction earthquakes and found that the asperities size were well predicted by this relationship. The EGFM is a well established methodology used in the field of earth sciences to estimate the ground motions. What the EGFM requires to estimate the ground motions according to the tectonic conditions of the region, are the fault parameters determination. Works like the one done by Irikura [15] address the study of the factors involved with the inner and outer fault parameters of the source to do more accurate estimations of ground motions. Those outer and inner fault parameters are estimated from the inversion of the waveform in studies of rupture processes using strong ground records. Using the results of many kinematic inversions, Somerville [6] obtained some relationships that synthesize the main characteristics of the earthquakes, stated as follows: (i) the seismic moment, (ii) the rupture area, (iii) the slip average, (iv) the combined area of asperities, and (iv) the area of the largest asperity, among others. In this study we are using these relationships because of their great utility in the estimation of seismic ground motions. The above statement provides an efficient way to work when a limited number of parameters to be considered in numerical simulations is available.

Finally from both estimations; it is said the PGAs, and the curves obtained with the two GMPE here used: (Ordaz [16], and Young's [17]), an trial and error iterative process of residuals minimization was conducted to identify the result that in the statistical sense better matched with the two GMPE here used.

The main contribution of this method is that it reflects a model that considers the source, the path, and the site effects. Another important contribution of the method is that reliable estimations about the energy distribution can be achieved in the high frequency band (between 0.1- and up to 10-Hz). This frequency range is of engineering interest because of the following reasons: (i) Many structures, including tall buildings and long bridges have their natural frequencies in the above frequency range, and (ii) 8 of the 10 major cities of this state are located in the sedimentary basins of the Colima graben and could amplify the ground motions in the frequency range of 0.1 to 10 Hz. It is therefore important to investigate how ground motions up to 10 Hz are generated from great subduction-zone earthquakes. This kind of investigations play a vital role in the effort to propose an scenario of strong ground motions from future large subduction earthquakes in the area in study and to evaluate the performance of structures subject to ground motions.

2. TECTONIC

The Colima state is located in Mexico's Pacific coast. The tectonic of the region is complex, in which the Rivera, the Cocos, and the North American plates converge. In addition to the above, the existence of a microplate has also been proposed by DeMets and Stein [18], and Bandy [19]. There are significant changes in the parameters of the subduction process along the subduction zone on the Pacific coast of Mexico, which has been divided in four sections by Pardo and Suárez [8]. Although the dip of the interplate contact geometry is constant to a depth of 30 km, lateral changes in the dip of the subducted plate are observed once it is decoupled from the overriding plate. In front of the Jalisco block, the Rivera plate has a dip of 45° and its subduction rate below the North American Plate is estimated to be from 2 to 5 cm/yr. The Cocos plate below Colima shows a similar dip to that of the Rivera but the subduction rate below the North American Plate is estimated to be from 4 to 6 cm/yr. To the south, the dip of the Cocos plate decreases gradually and is almost sub-horizontal at Guerrero (where it subducts with a velocity from 6 to 7 cm/yr) before increasing again farther south to the large values observed in Central America. Pardo and Suárez [8] explained the observed no parallelism between the volcanic belt and the subduction zone by these large lateral variations.

3. DATA

To simulate acceleration time histories of the Tecomán earthquake, we used two sources of data. Firstly, records of the 13 August 2006 earthquake. These records were obtained from previous temporal campaign in that area and from the support of other institutions with permanent instrumentation in that zone. Its epicenter, focal mechanism, and seismic moment were obtained from Centroid Moment Tensor Project [20]. The location of instruments that recorded this event (figure 1) is next described. The instruments were from permanent seismic networks: 15 Etna episensor wideband accelerographs from d.c. to 200 Hz at 200 samples per second from the national accelerations network of Instituto de Ingeniería (IINGEN) of Universidad Nacional Autonoma de Mexico (UNAM); two Guralp CMG40T-DM24 flat response wideband velocity type seismographs from 0.5 to 100 Hz at 100 samples per second from the network Red Sismica del Estado de Colima (RESCO). Secondly, data from temporal networks installed in the region as part of this project as follow: (i) four Altus Etna wideband accelerographs from d.c. to 100 Hz at 100 samples per second, four (ii) Geosig strong-motion recorder model 18 with analogue-digital converter, wideband accelerometers from d.c. to 100 Hz recording at 100 samples per second. Because 2 of the 25 records used in this study were velocity records, it was necessary to transform them to acceleration. Also, it was necessary to remove the instrumental response of each of the different instruments.

4. METHOD

The method used to model the target event requires a small magnitude event (earthquake of 13 August 2006) called element event, with hypocenter in close proximity to the earthquake that we want to simulate. For this particular case, the magnitude, location, focal mechanism and source parameters of the element event was reported by Harvard CMT (Mw 5.3, 18.45°N latitude and -103.63°W longitude, depth 23.5 km, strike 38°, dip 23°, and slip 96°, seismic moment 1.12e24 dyne-cm). For the simulated event (target event) and taking in consideration that the area in study is the region between the limits of the rupture areas of the Tecoman (2003) and the 1973 earthquakes, the hipocentral location was proposed just inside of the area in study and near to element event (18.45°N latitude and -103.75°W longitude). Considering the rupture area of 1973 earthquake and area of remainder gap of Tecoman earthquake we proposed 70 km along strike of fault area. Along the dip, we propose 80 km considering an intermediate value of dip length of neighbors earthquakes. Based on the above considerations the proposed effective rupture area is of 5600 km^2. Using equation (1), Somervile [6] we estimate a seismic moment of 1.1091e27 dine-cm.

$$A = 5.2^{-15} * Mo^{2/3} \tag{1}$$

Where *A* is the rupture area and Mo is the seismic moment.

Using equation (2) by Kanamori [21], the maximum estimated M_W magnitude is 7.3.

$$M_w = (1/1.5)\log(M_o) - 10.73 \tag{2}$$

Where *Mw* is the moment magnitude and Mo is the seismic moment.

We use the relationships of Somerville [6] to characterize the source parameters as follows: (i) equation (3) to estimate the combined area of asperities (*A2*), (ii) equation (4) to estimate area of largest asperity (*A1*), (iii) equation (5) to estimate the hipocentral distance to center of closest asperity (*CA*), and (iv) equation (6) to estimate the rise time (*Rt*) that is related to seismic moment of a small earthquake. For the S-wave propagation velocity we used 3.4 km/s.

$$A2 = 1.21e - 15 * Mo^{2/3} \tag{3}$$

$$A1 = 8.87e - 16 * Mo^{2/3} \tag{4}$$

$$CA = 1.76e - 8 * M0^{1/3} \tag{5}$$

$$Rt = 1.79e - 9 * Mo^{1/3} \tag{6}$$

The fault plane was defined considering: an azimuth of 38°, a dip of 23° and a slip of 96°. These parameters were taken assuming that the mainshock will have the same focal mechanism as the element earthquake.

To estimate the number of sub-events, we applied the ω^{-2} spectral model, Aki [22], obtaining the number of sub-events necessary and estimate N^3 by using the relationship between seismic moments of the target event (M_0), and the element event (m_0) that is used as empirical Green's function. N^3 is equal to the number of sub-faults in direction of the strike (Nx), the dip (Nw) and the time (Nt).

The above description clearly states that it is necessary to find the parameter N, which will be used to scale the fault area for the event to simulate. Since it is divided into N x N subfaults, N^3 is obtained using the equation (7), and the relationship between these parameters is stated through equations (7), (8), and (9).

$$N^3 = N_x x\, N_w x\, N_t \tag{7}$$

$$\frac{\overline{U_0}}{\overline{u_0}} = \frac{M_0}{m_0} = N^3, \tag{8}$$

where $\bar{U}_0$ and $\bar{u}_0$ is the flat level of the displacement Fourier spectrum for the target and element events respectively. On the other hand M_o and m_o are the seismic moments of the target and element events respectively. The relationship for high frequency is given by:

where $\bar{A}_0$ and $\bar{a}_0$ are the flat level of the acceleration Fourier spectrum of the target and element events respectively.

$$\frac{\overline{A_0}}{\overline{a_0}} = \left(\frac{M_0}{m_0}\right)^{1/3} = N, \tag{9}$$

Then the synthetic motion of the target event $A(t)$, is given by the element event $a(t)$ using the equations 10 and 11.

$$A(t) = \sum_{i=1}^{Nx} \sum_{j=1}^{N_W} \left(\frac{r}{r_{ij}} \right) F(t - t_{ij}) * a(t) \tag{10}$$

$$F_{ij}(t - t_{ij}) = \delta(t - t_{ij}) + \frac{1}{n} \sum_{k=1}^{(N-1)n} \delta[t - t_{ij} - \frac{(k-1)\tau}{(N-1)n}] \tag{11}$$

where n' is an appropriate value to eliminate spurious periodicity, r is the distance from the station to the element event hypocenter, r_{ij} is the distance from the station to the sub-fault (i, j), t_{ij} is the sum of the delay times due to the rupture propagation and the differences of distances between the location of the element event and the location of the target event (i,j) at the observed site.

During the simulation process, we use the Somerville [6] relationships to assume and vary the inner and outer fault parameters in order to simulate acceleration records. Such parameters are: the rupture velocity, the rise time and the point where the rupture starts among others.

The PGAs were estimated for the three orthogonal components. On the other hand, earthquake magnitude, focal depth, hypocentral distance, and site characteristics (rock or soil) were the controlling parameters to estimate curves of ground motion by using the GMPE from Ordaz [16] and Young's [17] used in this study. To compare our results with respective GMPE the PGAs of two horizontal components was computed according with table 1.

Table 1. Computation of two horizontal components in the two GMPEs.

GMPE	Horizontal components	Equation
Youngs *et al.* (1997)	Geometric mean *	$A_G = \sqrt{\max\|a_1(t)\|_{fort} \max\|a_2(t)\|_{fort}}$
Ordaz *et al.* (1989)	Cuadratic mean	$A_{max} = \sqrt{\frac{A^2_N + A^2_E}{2}}$

* Douglas (2003)

After the above steps were completed, we proceeded to generate and compare the mean value of the residual between the PGAs and each one of the GMPEs by applying the definition of mean residuals as the weighted sum of the residuals of the logarithmic values between observed and estimated. The above step was applied to identify, in the statistical sense trough the estimation of the residuals, how realistic our PGA estimations are.

5. RESULTS AND DISCUSSION

We applied the EGFM to generate a lot of models using the Somerville [6] relations to characterize the parameters of the source. The total rupture area was 5,571.90 km^2 with an average slip of 3.2 m and a combined area of asperities of 1,269.11 km^2, in which the area of largest asperity was of 812.33 km^2. The hipocentral distance to the center of the closest asperity was estimate at 18 km. It should be pointed out that the Somerville [6] relationships do not define an azimuth to specify the location of asperity, this mean the largest asperity can be located within an azimuth of between 0° to 360°. Then in the modeled process the position of largest asperity was varied from 0 to 360 ^{o}.

The main goal of this study was in the context of obtaining the most probable scenario of the ground response in the major cities of region upon the occurrence of a M_W 7.3 earthquake generated in the area in study. In this work, all the simulations were carried out to obtain a statistical sample that may represent the most probable scenarios of the ground response at the studied sites. For that purpose, the fault parameters and rupture process (i.e., (i) the azimuth of the closest asperity to hypocenter, (ii) the rise time, (iii) the rupture velocity, and (iv) the location of SMGA within the fault plane) were varied in an iterative process in order to generate a statistical sample of the most probable ground response scenario.

According to Somerville [6] an earthquake with a moment magnitude of 7.3 should have around 2.4 asperities. Based on the above, our procedure was divided in two stages. In the first stage, we used 2 SMGA and 3 SMGA in the second stage. These SMGA were positioned inside of different points into the dislocation area adopting the Somerville [6] criteria. In order to evaluate which scenario is the most probable we compare the PGA obtained from each scenario to those predicted by two GMPE, from Ordaz [16] and from Young´s [17].

It should be pointed out that both GMPE were obtained considering earthquake data from subduction tectonic environments as follows: (i) Ordaz [16] used subduction earthquakes from the Mexican pacific coast, and (ii) Young's [17] subduction earthquakes from around the world. The process of finding the best fit of the simulated PGA with respect to Ordaz [16], and Young's [17], GMPEs was based on the smallest residual criteria between the simulated PGA and the estimated GMPEs. The above, allowed us to identify 2 different source models that provided PGA values that better matched with the used GMPE as follows: (i) the first one with 2 SMGA, and (ii) the second one with 3 SMGA.

In order to find a best residual in each interaction, we varied the position of SMGA, the rupture velocity, rise time, radiation pattern, and the size of SMGA changes in the strike or dip directions, according to the azimuth of the station or stations that had poor adjustment with GMPE. In the process of modeling we found little sensitivity of the synthetics to the rise time variations. On the other hand, we found high sensitivity of the synthetics to the rupture velocity variation, the size of the SMGA and its location inside of fault plane. The parameters with major weight in the modeled are the number, the size and the location of SMGA. The optimal model is a combination of all these parameters. The best model for each stage was determined by minimizing the residual between synthetic and observed PGA.

Figures 2a and 2b show the comparison between both GMPE versus our results for the lowest residual case of the source models (two SMGA). Figure 2a shows the Young's [17] GMPE curves for rock and soil.

Table 2. Shows that the mean residual for the 25 stations decrease when using source models with 2 SMGA instead of when using source models with 3 SMGA. Table 2 shows the comparison of residuals between the theoretical values of each GMPE and the PGAs for all 25 stations.

Numbers of SMGA	SMGA	Length (km)	Width (km)	Area (km²)	Total area	Vr Km/s	Residual with Young's GMPE (Rock)	Residual with Young's GMPE (Soil)	Residual with Ordaz GMPE (Rock)
2	1	24.94	32.06	799.51	1256.38	3.0	0.009	0.024	0.245
	2	21.37	21.37	456.86		2.9			
3	1	35.62	24.94	888.35	1281.76	2.3	0.011	0.101	0.252
	2	17.81	10.69	190.36		2.5			
	3	14.25	14.25	203.05		2.4			

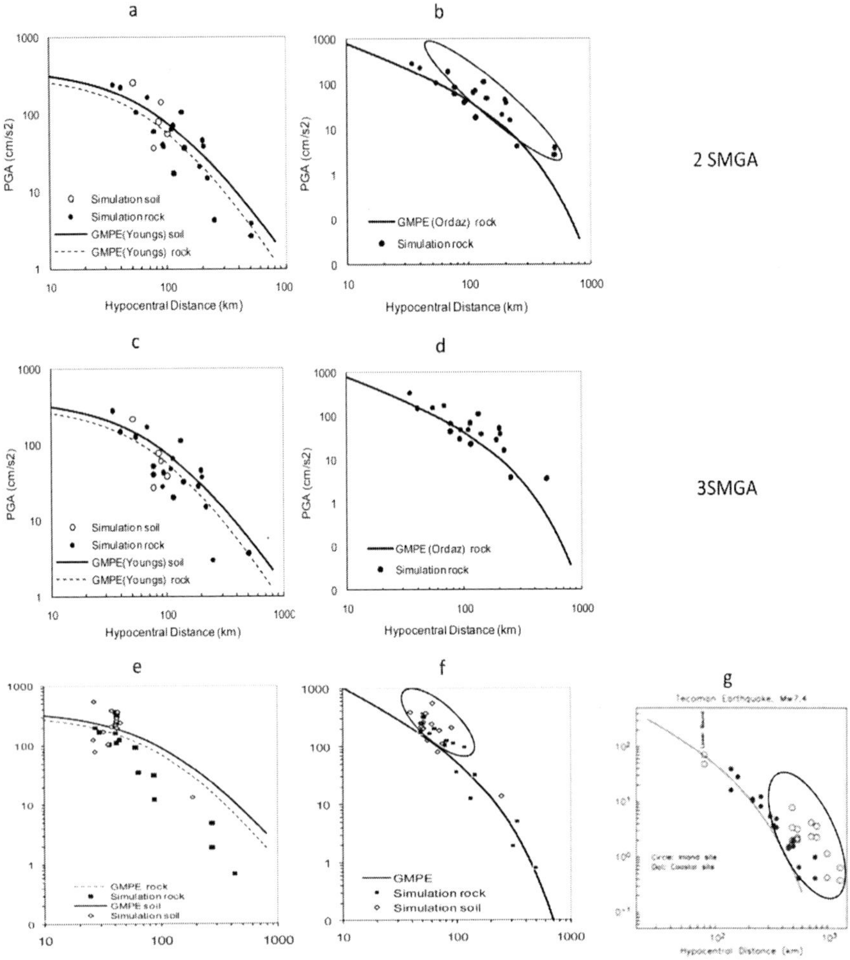

Figure 2. Comparison of synthetics PGA versus PGA from GMPE. In first row comparison of PGA from our simulated acceleration time histories using 2 SMGA versus PGA predicted using the following GMPE: (a) Young's [17], (b) Ordaz [16]. In second row comparison of PGA from our simulated acceleration time histories using 3 SMGA versus PGA predicted using the following GMPE: (c) Young's [17], (d) Ordaz [16]. In third row comparison of simulated PGA of Tecoman earthquake from Ramirez-Gaytán [23] versus PGA predicted using the following GMPE: (e) Young's [17], (f) Ordaz [16]. In (e) comparison of real PGA of the Tecoman earthquake using records of stations with distances larger than 50 km versus GMPE of Ordaz [16] from Singh [10].

Table 3. Relation of 25 stations where acceleration time histories was simulated, soil type, PGA, spectral acceleration corresponding to structural period of 0.1, 0.2, 0.3, 1.0, 2.0 and 3.0 s, EW component.

No	Station	Hipocentral distance	Soil type	PGA EW	Maximum acceleration (cm/s/s)					
					T 0.1	T 0.3	T 0.5	T 1	T 2	T 3
1	SJAL	34.54	Rock	457.66	819.32	1934.28	958.19	193.3	159.37	30.9
2	CEOR	40.07	Rock	125.38	218.08	265.65	328.99	178.02	273.65	57.36
3	BA5	51.22	Soil	157.32	230.26	402.48	428.66	150.87	325.69	54.23
4	COJU	54.02	Rock	201.97	248.95	484.43	190.27	65.25	39.12	8.63
5	MARU	68.11	Rock	194.29	419.83	559.86	627.33	261.75	180.06	29.4
7	TAPE	76.76	Rock	30	60.46	84.91	94.25	20.38	25.95	5.56
6	R15	76.78	Soil	27.33	45.06	47.41	51.59	16.83	7.49	2.14
8	MANZ	77.11	Rock	91.74	133.89	260.72	174.85	37.56	26.15	4.94
9	CAM	85.23	Soil	78.43	113.16	366.71	135.51	24.06	30.26	4.63
10	NAR	89.25	Soil	51.95	70.01	215.24	139.55	18.02	13.37	2.1
11	CEN	92.34	Rock	20.73	33.06	49.66	60.03	13.19	19.46	2.87
12	COMA	93.68	Rock	63.3	78.45	146.38	207.83	117.09	112.31	19.73
13	EZA	100.73	Soil	25.49	32.19	87.7	59.22	14.75	13.34	2.06
14	CIHU	108.51	Rock	42.84	47.23	91.04	213.08	26.49	12.49	2.56
15	EZ5	113.45	Rock	52.18	92.98	113.73	150.01	32.7	39.9	5.93
16	COLL	113.83	Rock	29.68	35	76.05	157.38	24.04	17.21	3.07
17	CALE	132.32	Soil	71.7	84.37	120.62	167.18	109.63	60.38	10.92
18	CDGU	140.21	Rock	50.83	57.72	177.79	172.63	43.18	15.1	3.34
19	VILE	188.48	Rock	31.74	32.76	44.1	98.98	60.66	48.56	8.31
20	NITA	199.77	Rock	68.46	75.31	188.1	193.54	120.04	80.13	13.41
21	CANA	203.51	Rock	31.33	43.4	85.63	77.88	60.92	46.25	8.62
22	URUA	217.28	Rock	20.51	21.37	34.56	69.81	38.08	56.54	9.33
23	GDLC	247.67	Rock	1.71	2.1	5.37	5.72	0.98	0.29	0.06
24	CUP	504.35	Rock	3.53	3.95	5.79	9.29	7.18	6.58	0.97
25	SCT	509.27	Soil	3.55	3.78	7.32	14.28	3.78	0.57	0.21

Our data showed three clusters that according to their hypocentral distances are distributed as follows: (i) The first group was distributed within the distance from 35 to 60 km, in this group 5 of the simulated PGA were located nearby of both curves; (ii) The second group is defined for distances range from 60 to 120 km, on this case the PGA are distributed almost evenly below the GMPE curves; (iii) Finally, the third group is defined for distances range from 120 to 500 km, on this particular situation the PGA values are over-estimated by the GMPE and show a clear tendency to attenuate faster than the pattern showed on the GMPE. Figure 2b shows the Ordaz [16] GMPE, the author uses thrust subduction earthquakes from

Mexico (such as event simulated in this study). In general the comparison shows that 90% our results are located above this curve.

For the GMPE of Ordaz [16], we compare only the 19 stations seated on rock. For the GMPE of Young's [17] we compare the 19 stations seated on rock, and 6 stations seated on soil sites (table 3), each of these groups with the respective curves for sites on rock and for sites on soil. From the modeling process of the target event, when three SMGA were used the lowest residuals we obtained between the PGA and the GMPE were: (i) 0.011 for Young's [17] GMPE (Rock), (ii) 0.101 Young's [17] GMPE (Soil), and (iii) 0.252 for Ordaz [16] GMPE for the sites shown in Figure 2c and 2d. The best fit between the estimated values of PGA and GMPE, estimated by means of the lowest residual, was obtained for the source model with two SMGA. The respective estimated residuals were: (i) 0.009 for Young's [17]GMPE (Rock), (ii) 0.024 for Young's [17] GMPE (Soil) and (iii) 0.245 for Ordaz [16].

Ramirez-Gaytán [23] simulate PGA and acceleration time histories for Tecoman earthquake located in the adjacent area at the event here simulated. The particularity and importance of this study is that the authors used observed strong motion records as a comparison reference to adjust the synthetics and found that from the comparison between PGA synthetics with respect to Young's [17] GMPE for rock and soil curves, and the Ordaz [16] GMPE curve (Figures 2e and 2f) behave very similar to the description previously provided for the same ranges of distances obtained in this study.

We made an additional comparison, for Tecoman earthquake Singh [10] used records of stations located at distances larger than 50 km and comparing the PGA versus Ordaz [16]. In figure 2g it can be seen that the behavior is the same: observed PGAs lie above the curve. The comparison made are important because in the first case Ramirez-Gaytán [23] generate PGA based in a previous model Ramirez-Gaytán [24] when using observed records to compare with synthetics. In the second case Singh [10] use real data, in both cases the results are very similar with the obtained in this study. All mentioned studies correspond at the same tectonic environment. Figure 2g show than for Singh [10] major PGA locate at distances above 120 km this explained because he use data from regional stations. In the case of Ramirez-Gaytan [23] major PGA are located at distances from 10 to 100 km this explain why authors only use data from local networks.

In this study, we compare our PGA's with those obtained by the GMPE of Ordaz [16] and because they used data from Mexico. However, all of these studies considered all of the data available at the time of their analyses. This means that the data used were essentially strong motion data from the southern part of the country Tejeda- Jácome and Chávez-García [25]. None of these studies included data from western Mexico, along the northern section of the subduction zone. It is uncertain that ground-motion prediction equations developed for Guerrero in a very different tectonic setting can be applied to Colima Tejeda-Jácome and Chávez-García [25]. For this reason, it is important to compare our results with GMPE of other parts of the world, such as the GMPE of Young's [17]. They used some similar parameters to those of the event simulated in this study (tectonic environment or hypocentral distance magnitude, etc.). This seems to be justified when we observed that the minor residual is reached when compared with the GMPE of Young's [17] with minor residual (0.009) than Ordaz [16] with residual of 0.245 for the model with best adjust (model with two SMGA). However, we expect that any of the two GMPEs compared in this paper satisfy the detailed similarities with the event simulated in this study for the Colima region. For this reason, our intention is to use local and world parameters in order to validate our results and to give a degree of confidence when applying this methodology to future earthquakes for determination of acceleration time histories and PGA.

The main contribution of the process detailed above is not obtaining PGA, for this case is sufficient to consult GMPE. The purpose of this paper is to prepare acceleration time histories to be used by structural engineers on the analysis and design of structures. In modern seismic design approaches the quality of a structural solution frequently depends on the detailed knowledge that designers have on the characteristics of the seismic ground motion that the structure will suffer at a site if an earthquake of a given magnitude occur at a given location. Figures 3a-3d show the acceleration time histories for the model with minor residual after applying the process previously described. The largest acceleration was obtained in the near-to-source station SJAL (rock site) with 0.47 g. It is important to point out that stations CUP and SCT (figure 3c) located in Mexico City with hipocentral distances larger than 470 km still produced considerable values of peak accelerations (3.53 and 3.55 gal) similar of those of the GDLC station (with 1.71 and 5.3 gal) located to 273.48 km from the epicenter. This is due to the fact that seismic waves are enormously amplified at lake-bed sites respect to hill-zone sites in Mexico City, although it has been suggested that even hill-zone sites suffer amplification and Singh [26].

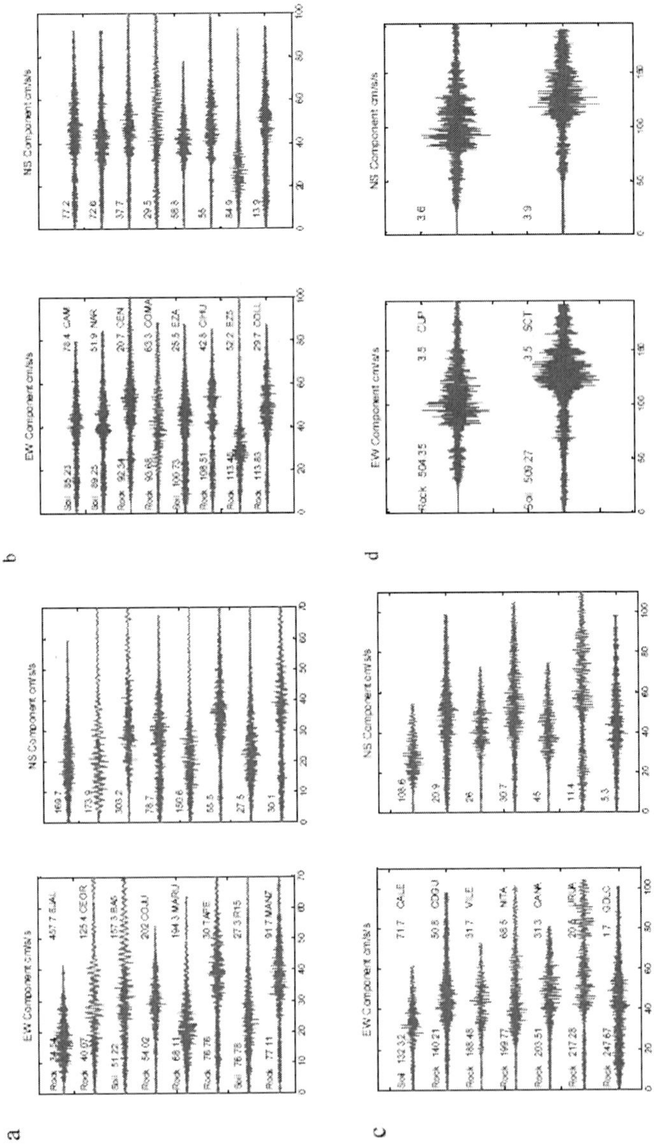

Figure 3. Simulated synthetic acceleration time histories (intense period) with next distances from hypocenter: (a) 34 to 78 km, (b) 85 to 114 km, (c) 132 to 248 km, (d) major to 500 km. In left column type of soil and epicentral distance are denoted above and left of each trace. PGA and code of station is denoted above and right of trace. In right column PGA is denote above and left of trace.

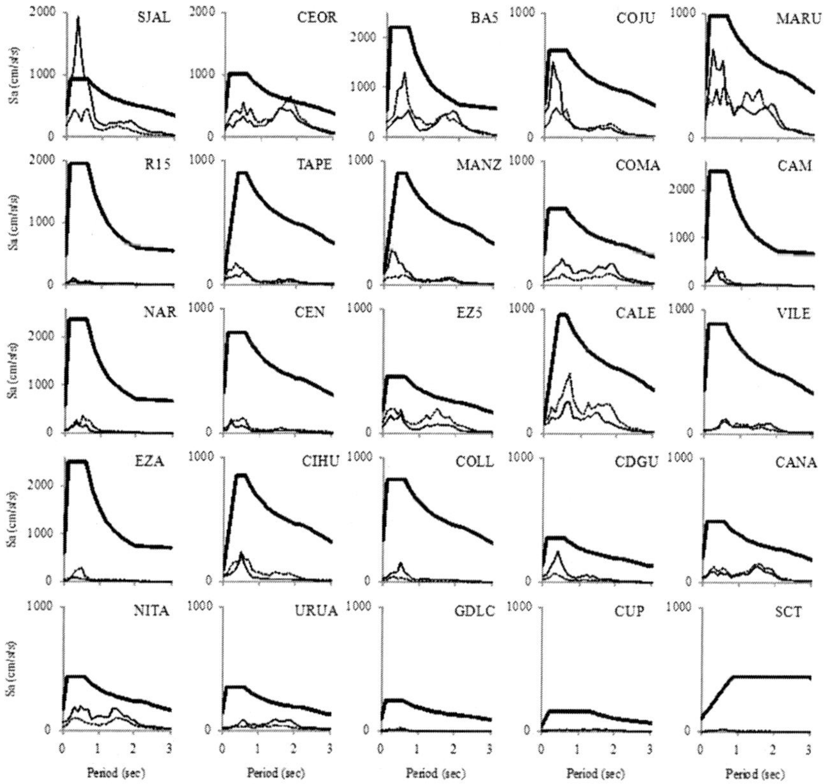

Figure 4. Comparison of the 25 response spectra ordered for hipocentral distance of the simulated event for east–west component (thin continuous line) and north–south component (discontinuous line) with the design spectra (dark continuous line). It can be observed that the response spectra obtained with EGFM is realistic.

On the other hand, with acceleration time histories it is possible to generate a response spectrum, which considers forces related to parameters of maximum response like spectral acceleration. Response spectra are essential for the seismic design, any effort to accurately predict these should be done. The synthetic acceleration response spectra for an equivalent viscous damping of 5 percent were calculated and compared with the elastic acceleration design spectra for structures of group B (standard occupancy) of the Manual of Civil Structures MOC-2008 [27], a model design code in Mexico and seismic provisions for current Mexico's Federal District Code NTCS-2004 [28], Sites CUP and SCT. In the MOC-2008

[27] code, seismic hazard in Mexico is defined as a continuum function where peak accelerations in rock are associated with return periods that were obtained using an optimization design criterion to define the seismic coefficients for the plateaus of the elastic design spectra for standard occupancy structures Tena-Colunga [29].

It can be also seen that only in 1 of 25 stations with distances comparable to the source dimensions the seismic ordinates are underestimated respect to the elastic design spectra. This station is SJAL located in rock site having seismic ordinate of 1.97 g in structural period of T= 0.3 sec. This is possibly because radiation patterns and source heterogeneity, which are taken into account in this work, causes spatial variations in ground motion around the fault Somerville [30]. These changes are very clear in the EGFM, but may be unnoticed in the GMPE. Therefore, obtaining future acceleration time histories and response spectra is a matter of essential interest for the present seismic design to achieve more efficient and safer structures.

6. CONCLUSIONS

We use the 25 records of August 13 2006 earthquake Mw 5.3 as element event to simulate strong ground motions for an eventual earthquake Mw 7.3 in the studied area. To reach this objective we integrate the advantages of three methodologies (EGFM, Somerville [6] relations and GMPE) to estimate the possible PGA, acceleration time histories and response spectra for an eventual earthquake in the studied area. We apply the empirical Green's function method (EGFM) whose main contribution is to reflect a model that considers the source, the path, and the site effects. In Mexico this method has been used to simulate an event that already occurred. In this study we applied it to predict some probable earthquake which may be expected in the region. To overcome the absence of observed records we made use of Somerville [6] relations to be able to make more accurate predictions of strong motions and two GMPE adequate for region to compare our results. The process of finding the best adjustment generated 2 different models (2 and 3 SMGA). This process of minimizing the residual between synthetics and observed PGA clearly shows that the mean residual for 25 stations is obtained when comparing with GMPE of Young's [17] and modeled with 2 SMGA. Ramirez-Gaytan [23] simulate PGA for Tecoman earthquake, whit difference that in this case the earthquake had occurred and exist observed records to compare and adjust synthetics, results are similar to those obtain in this study. Singh [10] comparing real PGA for Tecoman earthquake versus GMPE of Ordaz [16], results are similar to those obtain in this study. The purpose of this paper is to rescue the acceleration time histories of simulated event prepared to be used by structural engineers to analyze and design structures. Response spectrum show that for 1 of 25 stations (this station is near the source or with distance comparable with the source dimensions) the seismic ordinates are underestimated with the design spectra of the MOC-2008 [27] due possibly to radiation patterns and source heterogeneity, which is still to be confirmed by future records. For any practical evaluation of the seismic hazard in terms of response spectra is possible to integrate the advantages of three methodologies aboard in this study to estimate the possible PGA, acceleration time histories and response spectra for an eventual future earthquake.

7. ACKNOWLEDGEMENTS

We are grateful to Leonardo Alcantara Nolasco for providing acceleration records from the network of the Instituto de Ingeniería of UNAM, and Tonatiuh Dominguez for providing records of Red Sismica del Estado de Colima (RESCO).

REFERENES

1. Reyes A, Brune J, Lomnitz C, Source Mechanism and Aftershock Study of the Colima, Mexico Earthquake of January 30, 1973, Bulletin of Seismological Society of America, 69 (1979) 1819-1840.
2. Eissler H K, McNally K C, Seismicity and Tectonics of the Rivera plate and implications for the 1932 Jalisco Mexico earthquake, Journal of Geophysical Research, 89 (1984).
3. Singh S K, Astiz L, Havskov J, Seismic gaps and recurrence periods of large earthquakes along Mexican subduction zone: a reexamination, Bulletin of the Seismological Society of America, 71 (1981) 827 - 843.
4. Heaton T H, Kanamori H, Seismic potential associated with subduction in the northwestern United States, Bulletin of the Seismological Society of America, 74 (1984) 933 - 942.

5. Somerville P, Irikura K, Graves R, Sawada S, Wald D, Abrahamson N, Iwasaki W, Kagawa T, Smith N, Kowada A, Characterizing crustal earthquake slip models for the prediction of strong motion, Seismic Research Letters, 70 (1999).

6. Somerville P, Sato T, Toru I, Collins N, Dan K, Hiroyuki F, Characterizing subduction earthquake slip models for the prediction of strong motion, Proceedings of the Eleventh Symposium on Earthquake Engineering, 1 (2002) 163-166.

7. Nishenko SP, Singh KS, Conditional probabilities for the recurrence of large and great interplate earthquakes along the Mexican subduction zone, Bulletin of Seismological Society of America, 77 (1987) 2095-2114.

8. Pardo M, Suarez G, Shape of the subducted Rivera and Cocos plate in southern Mexico: Seismic and tectonics implications, Journal of Geophysical Research, 100 (1995) 357-374.

9. Minster J, Jordan T H, Molnar P, Haines E, Numerical modeling of instantaneous plate motions, Geophysics Journal, 36 (1974) 541-576.

10. Singh KS, Pacheco F, Alcantara L, Reyes G, Ordaz M, Iglesias A, Alcocer S, Gutierrez C, Valdez C, Kostoglodov V, Reyes C, Mikumo T, Quass R, Anderson R, A preliminary report on the Tecomán, México earthquake of 22 January 2003 (Mw 7.4) and its effects, Seismological Research Letters, 74 (2003) 279-289.

11. Quintanar L, Rodriguiez-Lozoya H, Ortega R, Gomez Gonzalez J, Dominguez T, Javier C, Alcantara L, Rebollar T, Source Characteristics of the 22 January 2003 Mw =7.5 Tecoman, Mexico, Earthquake:New Insights, in: Pure and Aplplied Geophysics, 2010.

12. Quintanar L, Rodriguiez-Lozoya H, Ortega R, Gomez Gonzalez J, Dominguez T, Javier C, Alcantara L, Rebollar T, Source Characteristics of the 22 January 2003 Mw =7.5 Tecoman, Mexico, Earthquake:New Insights, Pure and Aplplied Geophysics, 168 (201) 1339 - 1353.

13. Irikura Kojiro, Prediction of strong accelerations motions using empirical Green's function, Procedures of 7 th Japan Earthquake Engineering Symposium, (1986).

14. Aguirre J, Irikura K, Source Characterization of mexican Subduction Earthquakes from Acceleration Source Spectra for the Prediction of

Strong Ground Motions, Bulletin of Seismological Society of America, 97 (2007) 1960-1969.

15. Irikura K, Miyake T, Kamae K, Kawabe H, Dalguer LA, Recipe for predicting strong ground motion from future earthquakes, in: Proceedings of the 13th World Conference on Earthquake Engineering, available on CD-ROM, 2004.

16. Ordaz M, Jara J, Singh KS, Riesgo sísmico y espectros de diseño en el estado de Guerrero, Memorias del VIII Congreso Nacional de Ingenieria Sismica, (1989) D40-D56.

17. Youngs R, Chiou J, Silva J, Humphrey R, Strong ground motion attenuation relationships for subduction zone earthquakes, Bulletin of Seismological Society of America, 68 (1997) 58-73.

18. DeMets C, Stein S, Present-day kinematics of the Rivera plate and implications for tectonics of southwestern Mexico, Journal of Geophysical Research, 95 (1990) 21931-21948.

19. Bandy W, Mortera C, Urrutia J, Hilde T, The subducted rivera Cocos plate boundary: where is it, what is it, and is its relationship to the Colima rift?, Geophysical Research Letters, 22 (1995) 3075-3078.

20. Centroid Moment Tensor (CMT) Catalog, www.seismology.harvard.edu/CMTsearch.html (last access August 25 of 2008), in.

21. Kanamori H, The energy released in great earthquakes, Journal of Geophysical Research, 82 (1997) 2981-2987.

22. Aki K, Scaling law of seismic spectra, Journal of Geophysical Research, 72 (1967) 5359-5376.

23. Ramírez-Gaytán A, Aguirre GJ, Huerta CI, Simulation of Accelerograms, Peak Ground Accelerations, and MMI for the Tecomán earthquake of 21 January 2003, Bulletin of Seismological Society of America, 100 (2010).

24. Ramírez-Gaytán A, Aguirre GJ, Huerta CI, Tecomán Earthquake: Physical Implications of Seismic Source Modeling, Applying the Empirical Greens Function Method, and Evidence of non-Linear Behavior of Ground, ISET Journal of Earthquake Technology, 47 (2011) 1-23.

25. Jacome T, Chavez-Garcia FJ, Empirical ground motions estimations in Colima from weak motions records, ISET Journal of Earthquake Technology, 44 (2007) 409-419.

26. Ordaz M, Singh KS, Source spectra and spectral attenuation of seismic waves from Mexican earthquakes, and evidence of amplification in the hill zone of Mexico City, Bulletin of Seismological Society of America, 82 (1992) 24-43.

27. Comision Federal de Electricidad, Manual de Diseno de Obras Civiles 2008, Diseño por sismo, in: Instituto de Investigaciones Electricas (Ed.), 2008.

28. Reglamento de construcciones del Gobierno del Distrito Federal, NTCS-2004 Normas tecnicas complementarias para diseño por sismo, in: Gaceta Oficial del Departamento del Distrito Federal (Ed.), 2004.

29. Tena Colunga A, Mena Hernandez U, Perez -Rocha LE, Aviles J, Ordaz M, Vilar J, Updated Seismic Design Guidelines for Model Building Code of Mexico, Earthquake Spectra, 25 (2009) 869-898.

30. Somerville P, Smith N, Graves R, Modification of empirical strong ground motion attenuation relations to include the amplitude and duration effects of rupture directivity, Seismological Research Letters, 68 (1997) 199-222.

6

CHAPTER

GEOTECHNICAL, MINERALOGICAL AND CHEMICAL CHARACTERIZATION OF THE MISSOLE II CLAYEY MATERIALS OF DOUALA SUB-BASIN (CAMEROON) FOR CONSTRUCTION MATERIALS

Elisabeth Olivia Logmo[1], Gilbert François Ngon Ngon[1*], Williams Samba[1], Michel Bertrand Mbog[2], Jacques Etame[1]

[1]*Laboratory of Subsurface, Department of Earth Sciences, Faculty of Science, University of Douala, Douala, Cameroon*

[2]*Laboratory of Environmental Geology, Department of Earth Sciences, Faculty of Science, University of Dschang, Dschang, Cameroon*

ABSTRACT

Geotechnical tests conducted on clayey materials of Missole II, Douala sub-basin of Cameroon showed that these materials present: fines particles (55 to 78 wt.%), sand (22 to 44 wt.%), and plasticity index of 13.8 to 21.6%. The X-ray diffraction (XRD) and the chemical analysis revealed a kaolinite amount of 46 to 56 wt.%, 19 to 27 wt.% of illite, 12 to 19 wt.% of quartz, 3 to 5 wt.% of goethite, 2 to 5 wt.% of hematite, 1.5 to 5 wt.% of anatase, 2 to

3 wt.% of feldspar-K with 52.87 to 63.11 wt.% of SiO_2, 18.08 to 24.31 wt.% of Al_2O_3, 3.28 to 11.45 wt.% of Fe_2O_3 and a small content of bases (<2 wt.%). The results of geotechnical tests combined to those of the XRD and the chemical analysis showed that the Missole II clayey materials are suitable for the manufacture of bricks, tiles and sandstones.

Keywords: Cameroon; Construction Materials; Geotechnical Tests; Mineralogical and Chemical Analyses; Missole II

1. INTRODUCTION

Clay-rich materials are intensively used in the manufacturing of ceramics and as construction materials. However, the clayey deposit types are known as sedimentary, alluvial, and residual. A good knowledge of their occurrences, quantity and properties is required for their efficient exploitation.

In the Douala sedimentary sub-basin (South-Cameroon, Central Africa), some studies are done concerning the stratigraphic and tectonic evolution including [1-13]. Others studies are done to set up industrial units for manufacturing construction materials and ceramics [14] or concerning the mineralogical and chemical or thermal characteristics of the clay sediments [15-21].

In the Missole II area (Douala sub-basin), a geological study is carried out to locate and describe the clayey material outcrops and their provenance [22-24]. Also, some authors showed the possibility to obtain good ceramic building materials by mixing silica, feldspars, and kaolinitic and illitic clay of the Missole II area [25]. However, despite the preliminary works done on the field to describe clayey materials and to determine their sedimentation evolution, no further physical, mineralogical and chemical study is carried out to show their characteristics in order to their applications.

To that effect, the objective of this study is to associate the geotechnical characteristic to the mineralogical and chemical compositions of the clay occurrences of the Missole II deposit in order to evaluate its suitability for manufacturing of construction materials and ceramics.

2. GEOGRAPHIC AND GEOLOGICAL SETTING

Missole II is located on the Eastern part of the Douala sub-basin (Cameroon, Central Africa) between latitude 3°59' - 3°54'N and longitude 9°54' - 9°58'E. It is located within a humid equatorial climatic zone. Annual rainfall ranges between 3000 and 5000 mm, and the annual average temperature is 26°C. The vegetation is a dense rainforest transformed by human activities [26]. The geomorphology of the study area is a domain of the Cameroon coastal plain with low altitudes (40 - 120 m). The Missole II area shows hills with flat and sharp summits and is deeply dissected by V and U shaped valleys of MBongo, Bongougou, Missolo and Bongo the main rivers of the area. According to the geological map of SNH/UD report [12], the relative age of the Missole II sediments is Paleocene-Eocene corresponding to the N'Kapa Formation Figure 1.

The lithostratigraphy of Douala sub-basin is made of seven major Formations related to its geodynamic and sedimentary evolution [10-12]. 1) The synrift period represented by the Mundeck Formation (Aptian-Cenomanian) is discordant onto the Precambrian basement and consists of continental and fluvio-deltaic deposits, i.e., clays, coarse-grained sandstones, conglomerates. The postrift sequence includes; 2) the Logbadjeck Formation (Cenomanian-Campanian), discordant onto the Mundeck Formation and composed of micro conglomerates, sand, sandstone, limestone, and clay; 3) the Logbaba Formation (Maastrichtian), mainly composed of sandstone, sand and fossiliferous clay; 4) the N'kapa Formation (Paleocene-Eocene), rich in marl and clay with lenses of sand and fine to coarse-grained crumbly sandstone; 5) the Souellaba Formation (Oligocene) lying unconformably on N'kapa deposits and characterized by marl deposits with some interstratified lenses and sand channels; 6) the Matanda Formation (Miocene), dominated by deltaic facies interstratified with volcanoclasties layers; and 7) the Wouri Formation (Plio-Pleistocene) which consists of gravelly and sandy deposits with a clayey or kaolinic matrix.

3. RAW MATERIALS AND EXPERIMENTAL METHODS

The raw material used in this study comes from four representative clayey profiles of the Missole II area with three profiles of the interfluves along the Douala-Edéa road and one from the pit drilling on the lower slope of the valley. A geological survey shows different types of sediments with micro conglomerates, sandstones, fragments of ferruginous duricrusts and clays.

Clayey layers are overlain upwards by sandstones or micro conglomerates, ferruginous duricrusts and sandy-clays, and occupy the lower part of the profiles. Four clayey facies identified with different mixed textures like sandy-clay, clayey silt and silty-clay are mainly of sedimentary origin [23, 24]. The average thickness of the exploitable layers is 2.5 m. Two clayey samples collected from clay layer of each representative profile for mineralogical and chemical data are mixed to obtain average sample which served to realize the geotechnical analyses. A sufficient quantity of the single mixture of 2 to 3 kg of sediments is collected from a meter-long groove. The sample is analyzed for physical, mineralogy and chemistry.

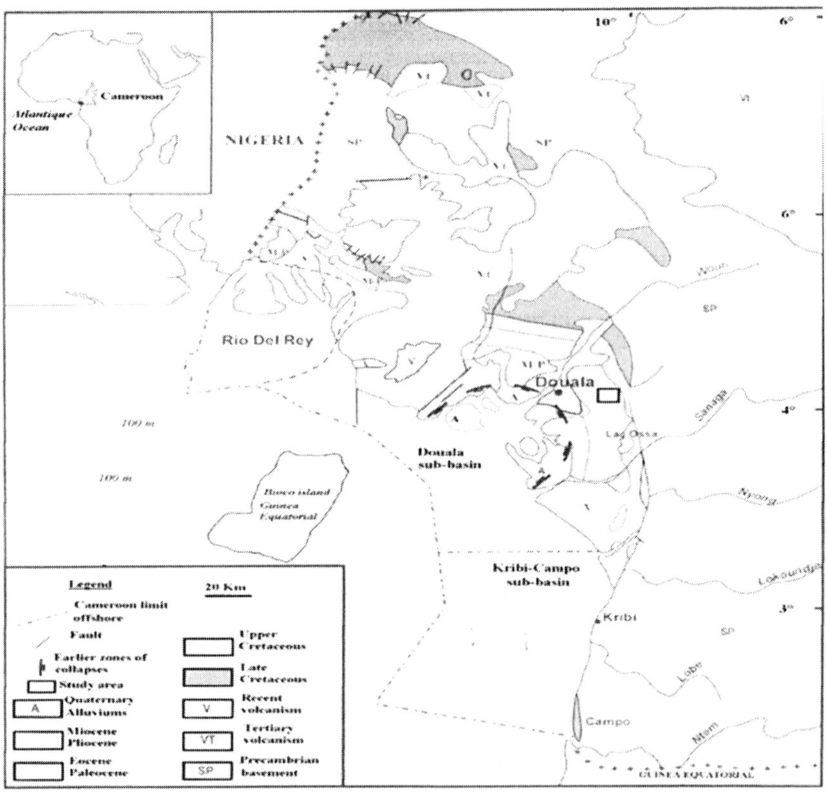

Figure 1. Geological sketch map of Cameroonian coastal basins (SNH/UD, 2005).

The particle size distribution has been achieved in two steps: 1) a conventional sieving for the 63 to 2000 μm fractions; 2) using a sedigraph 5000 in automatic procedure for clay and silt fractions. The liquid limit is measured by the method of the dish of Casagrande (wL) and the plastic limit by the method of the roller (wP). The blue methylene value (Vb) is determined on the total sample.

One hundred grams of each homogenized sample is grounded to −200 mesh (0.075 mm) in an agate mortar for chemical and XRD mineralogical study. Mineral identification is performed using a Setsys 2400 apparatus from SETARAM 85 equipped with a DSC 1500 heat system with Pt crucibles for thermal analysis, from room temperature up to 1100°C using a rate treatment of $10°C{\cdot}min^{-1}$ and alumina heat treated at 1500°C serving as reference material. For XRD, a Brünker diffractometer D8 ADVANCE with a copper source (λ = 1.5489 Å) is used on bulk and fine (<2 μm) samples, working under 40 kV and 40 mA. The exposure time for qualitative analysis is 2 h. Mineralogical phases are identified (JCPDS, 1998). Semi-quantitative analysis is performed [27]. For microscopic analysis, clay samples are examined with a scanning electron microscope (SEM) (Cambridge stereos can 200) coupled with an energy dispersive spectra microprobe (EDS). Homogenized powder of sediment sample is chemically analyzed for major and trace elements by ICP-AES after dissolution using acid digestion procedure with HF, HNO_3, and $HClO_4$. Classification of the clayey materials is performed by Autret (1983) method [28,29]. This method differentiates lateritic materials to non lateritic materials. It is based on the ratio S/R where, S is the ratio between SiO_2 concentration and the molar mass and R the ratio between Al_2O_3 plus Fe_2O_3 concentrations and their molar mass. In fact, for true lateritic material S/R is less than 1.33, for lateritic rocks it is 1.33 to 2, and for clay material it is more than 2.

4. RESULTS

4.1. Geotechnical Characteristics

The particle size distribution of the raw basic materials shows that the samples consist of 23 to 45 wt.% of sand, 17 to 33 wt.% of silts and 34 to 45 wt.% of clayey fractions. The geotechnical characteristics of the samples are reported in Table 1. These results are permitted to deduce that the class of these samples is in fine fractions [30]. The plasticity index (13.8% to 21.6%) is varied between 12 and 25, and shows that these materials are averagely plastic [31]. This plasticity is presumably a consequence of the average content of clayey minerals and quartz. The methylene blue value

(0.4 to 0.91 g/100g) of the concerned materials, which is corroborated with the plasticity index, may suggest the absence of swelling clay minerals.

4.2. Mineralogical Composition

4.2.1. DTA analysis

Figure 2 represents the DTA analysis curves obtained for the raw materials M2P3 and M2A3 at room temperature up to 1100°C using a rate treatment of 10°C·min^{-1}. Differential scanning calorimetric (DSC) observed on these DTA analysis curves shows endothermic pheno menon towards 110°C associated to the physisorbed water and another endothermic phenomenon towards 320°C due to the dehydroxylation of goethite. A specific endothermic transformation at 400°C - 600°C is due to the loss of structural hydroxyl groups of the kaolinite and an exothermic peak at 900°C - 1000°C due to the structural reorganization of the metakaolinite [32,33].

Table 1. Geotechnical characteristics of the samples.

Geotechnical characteristics	Results			
Average samples	M2A2	M2P3	M2P4	M2A3
Depth (cm)	600 - 750	700 - 800	700 - 800	50 - 150
Particles size distribution (wt.%)				
Sand (2000 - 63 µm)	45	31.3	44	22
Silts (63 - 2 µm)	21	25.5	17	33
Clays (<2 µm)	34	43.2	39	45
Atterberg limits				
Liquid limit, wL (%)	35.11	40.87	32.9	47.45
Plasticity limit, wP (%)	19.84	22.92	19.07	25.81
Plasticity index, PI (%)	15.27	17.95	13.83	21.64
Blue methylene value of the total sample (g/100g)	0.4	0.82	0.7	0.91

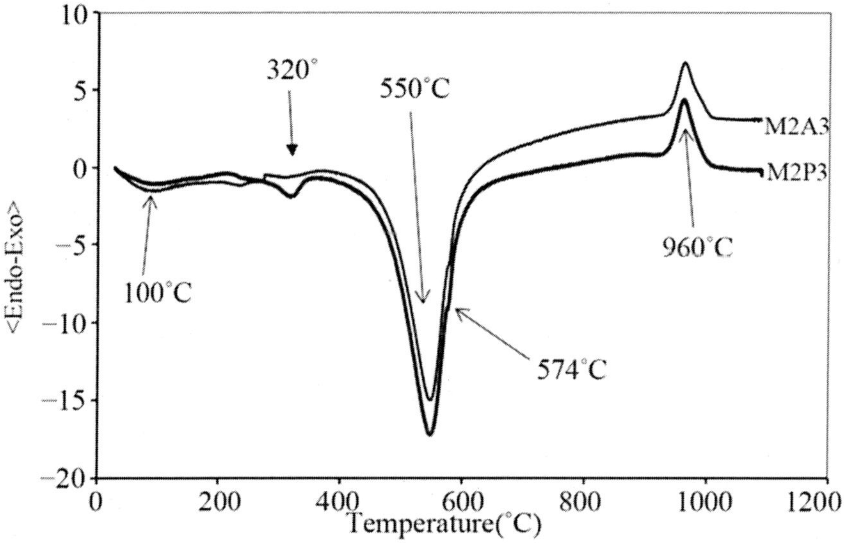

Figure 2. DTA curves of M2A3 and M2P3 materials for a heating rate of 10°C/min.

4.2.2. X-Ray Diffraction

The diffractograms of powder of the bulk samples ground until <80 μm revealed that clayey materials are kaolinitic and illitic and have permitted the detection of the following mineral phases: kaolinite, quartz, illite, goethite, hematite and accessory feldspar-K and anatase [24]. These mineral phases were usually present in theclayey materials of the Douala sub-basin [16,17,21].

The diffractograms of <2 μm fraction of Figure 3 also indicated the presence of the same mineral phases observed in the total samples with kaolinite, illite, quartz, goethite, hematite and accessory feldspar-K and anatase. Comparison of the minerals detected, these mineral phases are present in the fraction of the smaller and thicker granulometry.

The Table 2 gives the semi-quantitative mineralogical composition of the mineral phases of the samples. It reported an important proportion of kaolinite (46 to 56 wt.%), appreciable amounts of illite (19 to 27 wt.%) and quartz (12 to 19 wt.%), which are in relation with the value of the plasticity index (13.8 to 21.6%) and the methylene blue value (0.8 to 1.6 g/100g). The amounts of the other mineral phases are respectively goethite (3 to 5 wt.%), hematite (2 to 5 wt.%), anatase (1.5 to 5 wt.%) and feldspar-K (2 to 3 wt.%).

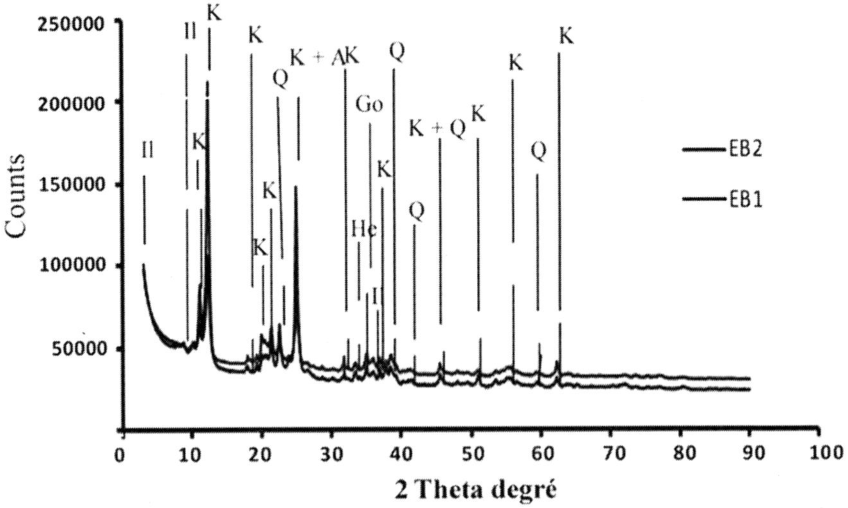
Counts
250000
200000
150000
100000
50000
0
10
20
30
40
50
60
70
80
90
100
2 Theta degré
EB2
EB1
Il
K
Q
K + A
Go
He
K + Q

(a)

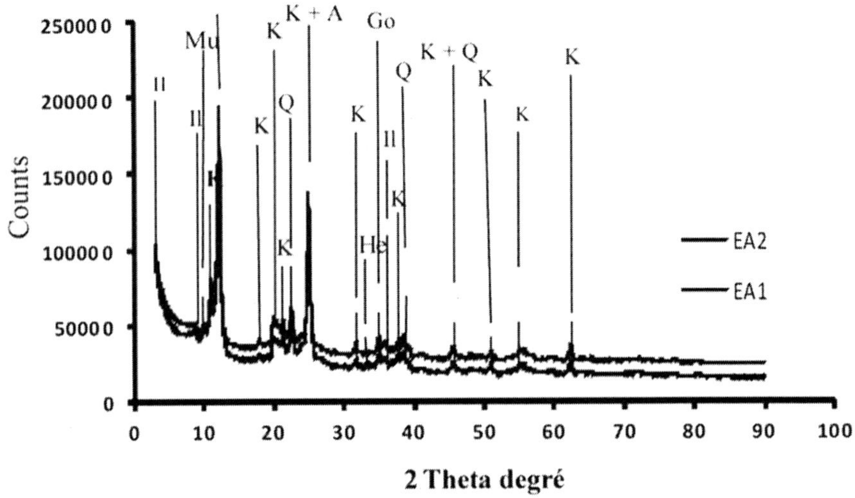
Counts
25000 0
20000 0
15000 0
10000 0
5000 0
0
10
20
30
40
50
60
70
80
90
100
2 Theta degré
EA2
EA1
Il
Mu
K
K + A
Q
Go
He
K + Q

(b)

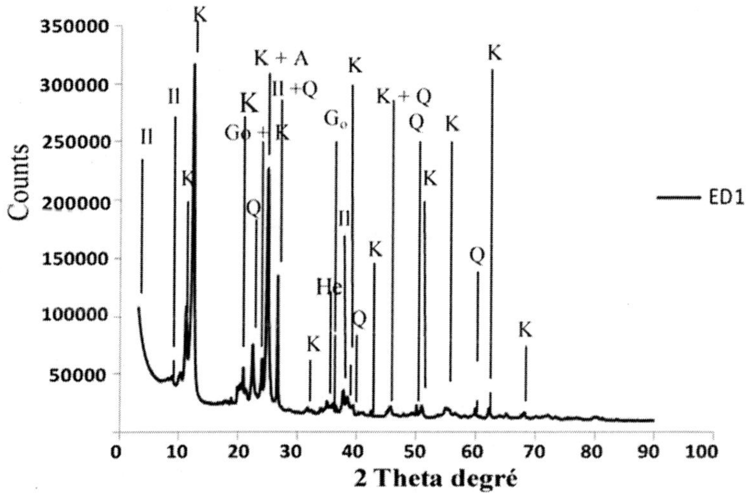

(c)

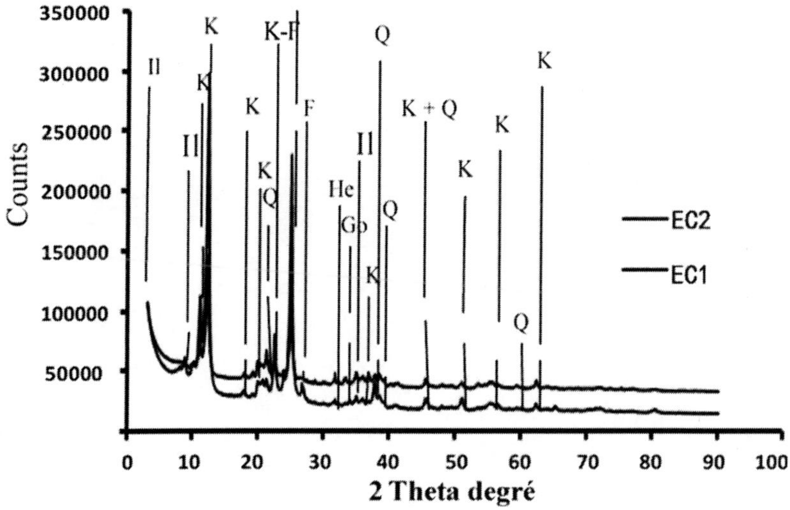

(d)

Figure 3. XRD patterns of <2 μm clay samples from Missolle II area of (a), (b), (c) and (d) profiles. A: Anatase; G: Goethite; He: Hematite; Il: Illite; K: Kaolinite; Q: Quartz; F: K-feldspar.

Table 2. Semi-quantitative mineralogical composition of the samples (wt.%).

	Profiles of the interfluves									Profiles of the pit	
Profiles	a			b			c			d	
Samples	**EA1**	**EA2**	**M2A2**	**EB1**	**EB2**	**M2P3**	**EC1**	**EC2**	**M2P4**	**ED1**	**M2A3**
Depth (cm)	600 - 700	700 - 750	600 - 750	700 - 750	700 - 750	750 - 800	700 - 750	750 - 800	700 - 800	50 - 100	50 - 150
Kaolinite	53.9	52.1	53.6	46.4	50.4	48.1	47.7	48.5	48.5	47.8	48.4
Illite	23.5	20.5	20.8	26.3	19.9	23.4	26.6	19.2	20.8	25.0	23.6
Quartz	16.1	15.0	17.1	18.8	17.7	18.2	12.5	17	16.9	13.7	16.1
Goethite	3.0	4.0	4.2	3.9	3.0	4.0	3.2	4.4	3.8	5.0	4.1
Hematite	-	2.2	1.6	-	3.0	2.1	3.2	4.4	3.6	2.9	3.2
Anatase	3.5	3.6	3.3	4.6	-	3.3	4.3	1.5	3.3	2.9	3.1
Feldspath-K	-	-	-	-	3.0	1.2	-	2.2	1.6	-	-

M2P3

M2A3

Figure 4. SEM images of M2P3 and M2A3 clayey materials of Missole II.

4.2.3. Microscopic Analysis

Figure 4 shows a SEM observation performed on samples of grey and mottled, with sandy-clay, silty-clay and clayey-silt texture. In clayey materials, piles of disordered kaolinites are observed with various sizes and irregular forms, which indicated that kaolinites of sedimentary clayey material of the Missole II area are poorly crystallized [34].

4.3. Chemical Composition

The chemical composition of the samples is given on Table 3. The results of the chemical analysis shows the presence of important amounts of SiO_2 (52.87 to 63.11 wt.%), associated with an appreciable amounts of Al_2O_3 (18.08 to 24.31 wt.%) and Fe_2O_3 (3.28 to 11.45 wt.%) and a small content of bases (<2 wt.%, CaO + MgO + Na_2O + K_2O).

These results show that quartz, alumino-silicates com pounds and iron minerals are predominate in the samples. The weak content of oxide calcium or potassium has proven that the calcitic and potassic minerals like calcite or feldspar are absent or are in a weak proportion. With S/R more than 2, the analyzed samples are clays [28].

Table 3. Chemical composition of the sample.

		Profiles of the interfluves									Profiles of the pit	
Profiles			a			b			c		d	
Samples	**DL**	**EA1**	**EA2**	**M2A2**	**EB1**	**EB2**	**M2P3**	**EC1**	**EC2**	**M2P4**	**ED1**	**M2A3**
Depth (cm)		600 - 700	700 - 750	600 - 750	700 - 750	700 - 750	750 - 800	700 - 750	750 - 800	700 - 800	50 - 100	50 - 150
SiO_2	0.01	63.11	56.85	58.01	54.97	57.24	56.23	52.87	54.02	54.08	63.1	62.84
Al_2O_3	0.01	18.94	24.31	23.06	22.16	19.1	20.87	23.24	20.59	21.01	18.08	18.61
Fe_2O_3	0.01	3.28	3.45	3.39	5.96	7.54	8.02	3.51	11.45	8.89	4.0	3.54
CaO	0.01	0.04	0.02	0.02	0.02	0.2	0.02	0.02	0.02	0.02	0.02	0.02
MgO	0.01	0.1	0.1	0.1	0.13	0.08	0.1	0.13	0.08	0.12	0.1	0.1
Na_2O	0.01	0.06	0.06	0.06	0.1	0.03	0.1	0.06	0.06	0.06	0.06	0.06
K_2O	0.01	0.43	0.4	0.49	0.77	0.21	0.3	0.47	0.47	0.47	0.6	0.7
TiO_2	0.01	1.02	1.38	1.42	0.99	1.1	1.0	1.2	1.2	1.2	0.78	0.83
MnO	0.01	0.01	0.01	0.01	0.01	0.01	0.01	0.01	0.01	0.01	0.01	0.01
P_2O_5	0.01	0.02	0.04	0.04	0.04	0.04	0.03	0.02	0.05	0.05	0.03	0.03
Cr_2O_3	0.01	0.01	0.01	0.01	0.01	0.02	0.01	0.01	0.02	0.01	0.01	0.01
LOI		13.02	13.12	13.29	14.21	13.85	13.21	17.38	11.81	14.01	14.4	12.34
Total		99.94	99.75	99.9	99.37	99.22	99.9	99.29	99.79	99.93	99.25	99.34

DL: Detection limit; LOI: Loss on ignition.

5. DISCUSSION

The particle size in this study revealed texture classes: silty-clay, clayey-silt and sandy-clay described in many deposits of the Douala sub-basin [17,21]. These classes characterized very sticky and plastic materials in accordance with some field characteristics such as smoothness, stickiness and plasticity [35]. Indeed, the fine particles and especially the very fine clay minerals determine the plasticity of an earthly material [32]. The plasticity index (13.8% to 21.6%) of the above clayey samples is supported by the high amounts of fine particles (silt + clay) and fine clay minerals. The amount of clay fraction is used to determine the manufactured products as is shown in Table 4 [36]. However, particle size and plasticity index are amongst the main geotechnical characteristics used to determine a convenient choice of construction material. In this case, taking into account of the amounts of sand (22 to 44 wt.%), fine particles notably the clay fractions (34 to 45 wt.%), and plasticity index, the Missole II clayey materials are convenient to the manufacturing of bricks, tiles and sandstones [37-39].

The mineralogical composition and geotechnical characteristics of a construction material must be known simultaneously in order to make a suitable choice for this raw basic material. It is well known that clay minerals have detrimental effects on the geotechnical properties; since some swelling clay minerals such as smectites absorb more water than the other (e.g. kaolinite). The volumetric shrinkage which increases with swelling clay minerals such as smectites constitutes an important parameter to take into account for civil public engineering [29,40]. The amount of clay minerals and especially swelling clay minerals contribute to the increasing of volumetric shrinkage and the swelling phenomena. Some authors joined the swelling potential and/or volumetric shrinkage to two parameters [36]. A first ordering is based on the plasticity index and the percentage of fine clayey fraction (<2 μm size). A second method proposes to use the liquid limit and the percentage of fine particles which size is less than 74 μm [41]. In this case, the studied materials, which have plasticity index between 12% and 25%, 34 and 45 wt.% of clayey fraction (<2 μm size), 54 and 78 wt% of fine fraction (<80 μm size) and liquid limit between 32% and 47% are very low to moderate swelling materials as is indicated in Table 5. This observation is corroborated with the mineralogical composition which shows that fine fraction is mainly composed of kaolinite and few illites, as well as smectites are absent.

Table 4. Relation between particles size and type of ceramic product to realize (after Cere and Mazel, 1993).

$\phi < 2$ µm (wt.%)	Ceramic products to realize
5 - 25	Full bricks
25 - 35	Perforated bricks
35 - 45	Hollow bricks, drains
40 - 50	Tiles, hordes

Table 5. Swelling potential of soils according to their plasticity index, liquid limit and respectively their <2 and <74 µm fractions.

PI (%)	% <2 µm	% <74 µm	wL (%)	Swelling potential
>35	>95	>95	>60	Very high
22 - 35	60 - 95	60 - 95	40 - 60	High
18 - 22	30 - 60	30 - 60	30 - 40	Moderate
<18	<30	<30	<30	Low

The geotechnical and mechanical characteristics of clayey materials are essentially depending on their chemical and mineralogical compositions as well as of the minerals distribution [37,39]. In this study, the clayey samples mainly constituted of poorly crystallized kaolinite (46 to 56 wt.%), which is in agreement with the loss on ignition (12% to 17%), illite (19 to 27 wt.%) and quartz (12 to 19 wt.%), are adequate for the composition of ceramic pastry [42] and for the manufacturing of bricks and tiles [14,37]. Also, clayey materials are siliceous (SiO_2 > 60 wt.%), with their alumina (Al_2O_3) content less than 35 wt.% and appreciable amount of iron (<10 wt.% with the exception of EC2 sample) is suitable for the manufacturing of tiles and bricks [41]. However, the weakest amounts of bases <2 wt.% (oxides of calcium, magnesium, sodium and potassium) which are the smelting or cimentitious compound corroborated with the flexural strength less than 10 MPa of the Missole II clayey material [25].

6. CONCLUSIONS

The following conclusions can be drawn out from the present work:

1) The geotechnical, mineralogical and chemical results show that the clayey materials of the Missole II, Douala sub-basin of Cameroon are mainly constituted of sand (23 to 45 wt.%), silts (17 to 33 wt.%) and clays (34 to 45 wt.%) with plasticity index of 13.8% to 21.6%.
2) Kaolinite, illite and quartz are the main minerals.
3) Clayey materials are also siliceous and aluminous, and have appreciable amount of iron in most clayey samples (Fe_2O_3 less than 10 wt%).
4) Based on the obtained geotechnical, mineralogical and chemical results, and the literature data, the studied materials are suitable for the manufacturing of bricks, tiles and sandstones.
5) These clayey materials present poor mechanical characteristics (feeble flexural strength and great fragility) partly linked to the chemical and mineralogical nature fundamentally complex of natural clays.

7. ACKNOWLEDGEMENTS

The authors acknowledge the financial support of the University of Douala and the geochemical analysis of Geo Labs (Geoscience Laboratories) in Canada. Many thanks are also given to Professor David Smith, Director of the GEMH laboratory at ENSCI and Professor Gisèle Lecomte of Centre Européen de la Céramique, Limoges, France, and to all of their collaborators.

REFERENCES

1. Y. C. Belmonte, "Stratigraphie du bassin sédimentaire du Cameroun," Proceeding on 2nd West African Micropaleontology Colloquium, Ibadan, 1-5 September 1966, pp. 7–24.

2. M. E. Brownfield and R. R. Charpentier, "Geology and Total Petroleum Systems of the West-Central Coastal Province (7203) West Africa," USGS: Geological Survey Bulletin, 2006, 2207-B 52 p.

3. E. Dartevelle and P. Brebion, "Mollusques fossiles du Crétacé de la Côte occidentale d'Afrique du Cameroun à l'Angola,"Annales du Musée Royal Congo Belge, Sciences Géologiques de Tervuren I-Gastéropodes, Vol. 8, No. 20, 1956, pp. 1-128.

4. E. Dartevelle, S. Freinex and J. Sornay, "Mollusques fossiles du Crétacé de la Côte occidentale d'Afrique du Cameroun à l'Angola,"Annales du Musée Royal Congo Belge, Sciences Géologiques de Tervuren II-Lamellibranches, Vol. 8, No. 20, 1957, pp. 1-271.

5. J. F. Dumort, "Identification par la Telédétection de l'Accident de la Sanaga (Cameroun)," Géodynamique, Vol. 1, No. 3, 1968, pp. 13-19.

6. J. B. Meyers, B. R. Rosendahl and H. Groschel-Becker, "Deep Penetrating MCS Imaging of the Rift-to-Drift Transition Offshore Douala and North Gabon Basins West Africa," Marine Petrology Geology, Vol. 13, No. 7, 1996, pp. 791-835.

7. D. Reyre, "Histoire géologique du bassin de Douala," In: D. Reyre, Ed., Symposium sur les Bassins Sédimentaires du Littoral Africain, Association du Service Géologique d'Afrique, IUGS, 1966, pp. 143-161.

8. P. R. N. Ngaha, "Contribution à l'Etude Géologique, Stratigraphique et Structurale de la Bordure du Bassin Atlantique du Cameroun," Thèse 3e Cycle, Université de Yaoundé, 1984.

9. P. Maurizot, A. Abessolo, J. L. Feybesse, V. Johana and P. Lecomte, "Synthèse des Travaux de 1978 a` 1986," Rapport 85CM066, 1986.

10. J. M. Regnoult, "Synthèse Géologique du Cameroun," D.M.G., Yaoundé, Cameroun, 1986.

11. F. R. Nguene, S. Tamfu, J. P. Loule and C. Ngassa, "Paleoenvironments of the Douala and Kribi/Campo Subbasins in Cameroon, West African," Colloque de Géologie Africaine, Libreville, Recueil des Communications, Géologie Africaine, 6-8 May 1991, 1992, pp. 129-139.

12. SNH/UD, "Stratigraphie Séquentielle et Tectonique des Dépôts Mésozoïques Synrifts du Bassin de Kribi/Campo," Rapport Non Publié, 2005.

13. C. S. Manga, "Stratigraphy, Structure and Prospectivity of the Southern Onshore Douala Basin Cameroon—Central Africa," In: M. J. Ntamak-Nida, G. E. Ekodeck and M. Guiraud, Eds., Cameroon and Neighboring Basins in the Gulf of Guinea (Petroleum Geology tectonics Geophysics Paleontology and Hydrogeology), African Geoscience, 2008, pp. 13-37.

14. P. M. Thibaut and P. Le Berre, "Recherche d'Argiles Pour Briques Dans la Région de Yaoundé, Douala et Edéa," Rapport 85CM065, 1985.

15. D. Njopwouo, "Minéralogie et Physico-Chimie des Argiles de Bomkoul et de Balengou (Cameroun). Utilisation Dans la Polymérisation du Styrène et Dans le Renforcement du Caoutchouc Naturel," Thèse d'Etat, Faculté des Sciences, Université de Yaoundé, 1984.

16. D. Njopwouo and R. Wandji, "Minéralogie de l'Argile Kaolinique de Bomkoul (Cameroun)," Revue de Sciences et Technique, Série des Sciences de la Terre, I 3-4, 1985, pp. 71-81.

17. D. Njopwouo and S. Kong, "Minéralogie de la Fraction Fine des Matériaux Argileux de Bomkoul et de Balengou (Cameroun)," Annales de la Faculté des Sciences, Série des Sciences Chimiques, I 1-2, 1986, pp. 17-31.

18. A. Elimbi and D. Njopwouo, "Firing Characteristics of Ceramics from the Bomkoul Kaolinite Clay Deposit (Cameroon)," Tile and Brick International, Vol. 18, No. 6, 2002, pp. 364-369.

19. A. M. M. Mpondo, "Cartographie des Affleurements de la Localité de Bomkoul (Sous-Bassin de Douala)," Université de Douala, Mémoire, 2010.

20. M. B. Mbog, "Etude Morphologique, Physico-Chimique et Minéralogique des Argiles de Bomkoul Dans le Sous-Bassin Sédimentaire de Douala-Cameroun," Université de Douala, Douala, 2010.

21. G. F. Ngon Ngon, J. Etame, M. J. Ntamak-Nida, M. B. Mbog, A. M. Maliengoue Mpondo, M. Gérard, R. Yongue-Fouateu and P. Bilong, "Geological Study of Sedimentary Clayey Materials of the Bomkoul Area in the Douala Region (Douala Sub-Basin, Cameroon) for the Ceramic Industry," Comptes Rendus Geoscience, Vol. 344, No. 6, 2012, pp. 366-376.

22. W. Samba, "Etude Morphologique, Géotechnique et Minéralogique des Argiles de Missole 2 Dans le SousBassin de Douala-Cameroun," Université de Douala, Douala, 2010.

23. E. O. Logmo, "Etude Géologique, Minéralogique et Géochimique des Argiles de Missole II Dans le Sous-Bassin de Douala (Cameroun)," Université de Douala, Douala, 2012.

24. G. F. Ngon Ngon, E. Bayiga, M. J. Ntamak-Nida, J. Etame, S. Noa Tang and R. Yongeu-Fouateu, "Trace Elements Geochemistry of Clay Deposits of Missole II from the Douala Sub-Basin in Cameroon (Central Africa): A Provenance Study," Sciences, Technologie et Développement, Vol. 13, No. 1, 2012, pp. 20-35.

25. G. F. Ngon Ngon, G. L. Lecomte Nana, R. Yongue Fouateu, G. Lecomte and P. Bilong, "Physicochemical and Mechanical Characterisation of Ceramic Materials Obtained from a Mixture of Silica, Feldspars and Clay Material of the Douala Region in Cameroon (Central Africa)," Advances in Ceramic Science and Engineering (ACSE), Vol., 2, No. 1, 2013, pp. 23-31.

26. R. Letouzey, "Atlas du Cameroun, Phytogéographie Camerounaise," Imprimerie Nationale Yaoundé, 1968.

27. A. K. Chakravorty and D. K. Ghosh, "Kaolinite-Mullite Reaction Series: The Development and Significance of a Binary Aluminosilicate Phase," Journal of the American Ceramic Society, Vol. 74, No. 6, 1991, pp. 1401-1406.

28. P. Autret, "Latérites et Graveleux Latéritiques," Laboratoire Central des Ponts et Chaussées, 1983.

29. M. Younoussa, "Etude Géotechnique, Chimique et Miné- ralogique de Matières Premières Argileuses et Latéritiques du Burkina Faso Améliorées aux Liants Hydrauliques: Application au Génie Civil (Bâtiment et Route)," Thèse, Université de Ouagadougou, 2008.

30. NF P94-057, "Analyse Granulométrique des Sols. Méthode par Sédimentation," AFNOR, 1992.

31. NF P94-051, "Détermination des Limites d'Atterberg," AFNOR, 1993.

32. C. A. Jouenne, "Traité de Céramiques et Matériaux Mineraux," Septima, Paris, 1990.

33. S. Lee, Y. J. Kim and H. S. Moon, "Phase Transformation Sequence from Kaolinite to Mullite Investigated by an Energy-Filtering

Transmission Electron Microscope," Journal of the American Ceramic Society, Vol. 82, No. 10, 1999, pp. 2841-2848.

34. M. W. Carty, "The Colloidal Nature of Kaolinite," The Bulletin of the American Ceramic Society, Vol. 78, 1999, pp. 72-76.

35. E. A. Fitzpatrick, "Soils-Their Formation, Classification and Distribution," Longman, Berlin, 1983.

36. L. Cere and F. Mazel, "Caractérisation d'Argiles," ENSCI Limoges, 1993.

37. Y. Beron and P. Le Berre, "Guide de Prospection des Matériaux," Manuels et Methodes du BRGM, No. 5, 1983.

38. V. Rigassi, "Bloc de Terre Comprimée," Manuel de Prospection, Vol. 1, Craterre EAG, 1995.

39. N. Française, "Sols: Reconnaissance et Essais. Description-Identification-Dénomination des Sols," XP P94-011, 1999.

40. A. A. Al-Rawas and M. Qamarouddin, "Construction Problems of Engineering Structures Founded on Expansive Soils and Rocks in Northern Oman," Building and Environment, Vol. 33, No. 2-3, 1998, pp. 159-171.

41. A. Djedid, A. Bekkouche and A. M. Aissa Mamoune, "Identification and Prediction of the Swelling Behavior of Some Soils from the Tlemcen Region of Algeria," Bulletin des Laboratoires des Ponts et Chaussées, Vol. 233, 2001, pp. 69-77.

42. O. Castelein, "Influence de la Vitesse du Traitement Thermique sur le Comportement d'un Kaolin: Application au Frottage Rapide," Thèse, Université de Limoges, 2000.

7

CHAPTER

CORRELATION OF STRUCTURAL SEISMIC DAMAGE WITH FUNDAMENTAL PERIOD OF RC BUILDINGS

Anastasia K. Eleftheriadou, Athanasios I. Karabinis

Laboratory of RC, Department of Civil Engineering, Democritus University of Thrace, Xanthi, Greece

ABSTRACT

The sufficient estimation of the natural period of vibration constitutes an essential step in earthquake design and assessment and its role in the development of seismic damage is investigated in the current research. The fundamental period is estimated for typical reinforced concrete building types, representative of the building stock of Southern Europe, according to existing relationships. The building typologies also represent groups of 180,945 existing damaged buildings of an observational database created after the Athens (7-9-1999) near field earthquake. The estimated fundamental periods are correlated to several degrees of the recorded damage. Important conclusions are drawn on the parameters (height, structural type, etc.) that influence the seismic response and the development of damage based on the wide database. After conducting a correlation analysis, noticeable is the difference between the seismic demand of the elastic spectrum of the first (1959), the contemporary

(2003) Greek Seismic Code and the values of peak ground accelerations of several Athens earthquake records. Moreover, PGAs in most records are often between the lower and the upper bound of the estimated fundamental periods for RC buildings with regular infills (n-normal) and with ground levels without infill panels (p-pilotis) regardless the height. A disparity is noticed when the estimated fundamental period is based on EC8 provisions for the considered as "high" buildings in S. Europe regarding the referring earthquake. The majority of buildings that developed several degree, type and extent of damage are considered of "low" height with estimated fundamental periods close to the PGA values of Athens earthquake ground motions. However, the developed damage was the result of the combination of parameters based on geological, tectonic and morphological characteristics of the affected area. In addition, a damage scale for the measurable recording, beyond the qualitative characterization of seismic damage in Greek post-earthquake surveys, is presented wherein the performance levels are defined according to the physical description of the seismic damage and, as well, in terms of structural and economic damage index.

Keywords: Seismic Damage; Period of Vibration; Fundamental Period; Damage Scale; Observational Damage Database; Athens Earthquake

1. INTRODUCTION

The reliable evaluation of a basic dynamic characteristic as the fundamental period is an essential step in estimating the seismic response both in seismic design and assessment. The period of vibration is mainly dependent on mass, stiffness and strength and consequently on all the factors which affect them (dimensions, section properties, irregularities, etc.). The period of vibration is dependent mostly on mass, stiffness and strength and on all the factors which affect the pre-mentioned characteristics such as building material and type, dimensions (in height and plan), morphology and irregularities, section properties, stiffness, cracking, etc. Many simple relationships have been proposed in an effort to determine the fundamental period of structures. These empirical formulas have been produced by fitting curves through regression analysis on the buildings periods of vibration measured from their seismic movements during several earthquakes. However, Goel and Chopra [1,2] noticed that the different soil conditions, seismicity and design and construction practices may affect the measured fundamental periods. Despite the fact that several parameters affect the period of vibration the empirical formulas are given as a function of height as it has been found to play the most important role [3,4].

In the current research the fundamental period of vibration is calculated for several reinforced concrete building types according to existing simple relationships. The typical building typologies are representative of the materials, the seismic codes and the construction techniques of Southern Europe. They also represent groups of existing damaged buildings in several degree, type and extent of a dataset derived from post-earthquake surveys which took place in an extended region of Attica in Greece. The damage database with thousands of recorded buildings created after the occurrence of the Athens (7-9- 1999) near field seismic event in an extended urban region [5-7]. The estimated fundamental period of typical reinforced concrete (RC) structures is correlated with the recorded seismic damage. The advantage of the research is that the empirical vulnerability assessment is the most realistic as it is based on a real experiment in a physical scale 1:1. Important conclusions are drawn on the parameters (height, structural type, etc.) that influence the seismic response and the development of damage based on the wide database. In addition, a damage scale for the measurable recording, beyond the pre-mentioned quailtative characterization of seismic damage in Greek postearthquake surveys, is presented wherein the performance levels are defined according to the physical description of the seismic damage and, as well, in terms of structural and economical damage index. A correlation is fulfilled between the estimated fundamental periods of RC buildings with the different damage states of the proposed scale, FEMA performance levels and the analysis results of the database.

2. ESTIMATION OF PERIOD OF VIBRATION FOR RC BUILDINGS

The reliable and sufficient estimation of the natural period of vibration of a RC building could play an essential role in the understanding of the global demands on the structure under an earthquake. In most design codes the estimation of the fundamental period is related to the building's height based on simple empirical relationships given the structural typology. These formulas have been derived from empirical data from existing buildings subjected to seismic actions. In the equivalent static method (linear) the prediction of the fundamental period of vibration in a simplified manner allows the calculation of the design base shear force which is then distributed along the height of the building in a linear manner. In the linear dynamic (or modal response spectrum) method an analytical model of the structure is performed (often using structural sections of reduced stiffness) in order to calculate the periods of vibration and a number of significant modes. The forces resulting from each mode are applied to the building using the appropriate modal shape and the seismic actions resulting from these forces are combined according to the provisions [8].

The force-based design produces conservative values of period which consequently leads to conservative estimates of base-shear force. As far as the assessment of building's response is concerned, the crucial parameter which is correlated to the damage development is the displacement [9] which may also be underestimated due to the conservative period-height estimation [3]. Recently, a deformationbased method was suggested [10] producing displacementperiod capacity curves for determining the non-linear demand at a given limit state using analytical relationships of displacement-height capacity curves and empirical relations of height-elastic period. This evaluated via analytical procedures period, which is larger than the elastic one, deals with the point of yielding and cracked stiffness of the structure and depends on the structural modeling assumptions [11]. It is well known that the stiffness degradation, due to the increase of the degree of damage, when a building is subjected to an earthquake also leads to a period increase.

2.1. RC Resisting Frames (MRF) Buildings

The first empirical formula in seismic design codes was presented in the USA building code ATC3-06 [12] and had the form of Equation (1):

$$T = C_t H^{3/4} \quad \text{or} \quad T = C_t H^{0.75} \tag{1}$$

where C_t is a regression coefficient calibrated in order to achieve the best fit to experimental data and H represented the height of the building in feet. The form of Equation (1) has theoretically derived from the Rayleigh's method by assuming that the equivalent static lateral forces are linearly distributed over the height of the building and the base shear force is assumed to be proportional to $1/T^{2/3}$. In ATC3-06, the value for the coefficient C_t in Equation (1) was proposed equal to 0.025 as a lower limit for evaluating the period of vibration of RC MRF buildings. In SEAOC-88 [13] it was found that a value of C_t = 0.03 was more appropriate for reinforced concrete buildings and has been adopted in many design codes (UBC-97, SEAOC-96, NEHRP-94, NBCC 2005 and Eurocode 8). Other values (0.035 or 0.04 in feet or 0.085 and 0.097, respectively in metres) for the coefficient C_t have also been recommended [8,14]. ASCE7-05 [15] adopted the period-height formula proposed by Goel and Chopra [1] who found that the best fit lower bound for reinforced concrete frames was given by the following expression:

$$T = 0.0466H^{0.9}\left(H \text{ in metres}\right) \tag{2}$$

In the European EC8 [16] the period-height expression for moment resisting frames adopted the same values by SEAOC '99:

$$T = 0.075H^{0.75} \tag{3}$$

The last has also been derived using Rayleigh's method and the height has been transformed from feet in metres ($0.03 \times 3.281^{0.75} = 0.073$).

2.2. RC Shear Walls (SW) Buildings

The period of vibration of these buildings is usually evaluated with good approximation as the period of an equivalent cantilever. In reinforced concrete shear walls buildings both the lateral load resistance and the lateral stiffness is mainly provided by shear walls. The first period-height relationship for RC Shear Walls (SW) buildings had the form of Equation (4):

$$T = \frac{C_t H}{\sqrt{D}} \tag{4}$$

where D is the dimension of the building at its base in the direction under consideration. In ATC3-06 [12] C_t was proposed as 0.09 with the dimensions measured in metres. This formula comes from the equation of the frequency of vibration of a cantilever (considering shear deformations only), with the thickness of the wall considered to be more or less constant and thus only the width/length of the building is an input parameter, as demonstrated in Equation (5) [8,17]:

$$T = 4\sqrt{\frac{m}{kG}}\frac{H}{\sqrt{A}} = \frac{aH}{\sqrt{A}} = \frac{aH}{\sqrt{Dt_w}} = \frac{a_1 H}{\sqrt{D}} = \frac{C_t H}{\sqrt{D}} \tag{5}$$

where m is the uniformly distributed mass per unit height, G is the shear elastic modulus, k is the shape factor to account for non uniform distribution of shear stresses for the shape of the transverse section (equal to 5/6 for a rectangular section), D is the length of the cantilever, A is the area of the transverse section expressed as the product of the thickness t_w and the plan dimension D of the equivalent cantilever parallel to the considered direction. Many seismic codes adopted the pre-demonstrated according to the building type. However, UBC-97 code [18] consisted in the use of empirical equations of the form of Equation (1) for shear wall buildings, where C_t was taken equal to 0.05 with the height measured in metres. Many engineers following Eurocode 8 would probably also use Equation (1) with C_t equal to 0.075 for all RC buildings regardless of the details of the masonry infills. Taking into consideration that many European reinforced concrete buildings are constructed with stiff masonry infill panels which are often not isolated from the RC frame, the period of vibration is probably being overestimated by the designers and hence the forces are subsequently being underestimated.

In SEAOC-96 [19] and UBC-97 [18] the proposed value of C_t regarding the shear walls buildings of Equation (1) is given following the expressions:

$$C_t = \frac{0.1}{\sqrt{A_c}} \tag{6}$$

and

$$A_c = \sum_{i=1}^{NW} A_i \left[0.2 + \left(\frac{D_i}{H} \right) \right] \text{ with } \frac{D_i}{H} \leq 0.9 \tag{7}$$

where A_c, the combined effective area (in square feet) of the shear walls, D_i is the dimension in the direction under consideration (in feet) of the ith shear wall in the first storey of the structure and NW is the number of shear walls.

In ASCE7-05 [15] the relative period-height equation is expressed by Equation (8):

$$T = \frac{0.0019}{\sqrt{C_w}} H \tag{8}$$

with

$$C_w = \frac{100}{A_B}\sum_{i=1}^{NW}\frac{A_i}{\left[1+0.83\left(\frac{h_i}{D_i}\right)^2\right]} \tag{9}$$

Finally, in Eurocode 8 [19] the formula has also adopted (as in MRF buildings) the same values by SEAOC '99 for structures with concrete or masonry shear walls:

$$T = 0.05H^{0.75} \tag{10}$$

where H is the height measured from base in metres.

Simplified expressions are used for the estimation of the fundamental period of typical Greek RC buildings with several heights in order to be afterwards correlated with the spectral accelerations of the 7th of September Parnitha's seismic event. Both the mean period of T_m (50% reliability) and the characteristic period T_R (95% reliability) value have been estimated of the fundamental period of RC buildings with regular infills (uniform stiffness/n-normal) and with ground level without infill panels (p-pilotis) [20]. The following experimental formulas refer to typical buildings in Southern Europe and they are used in the estimation of the above mentioned fundamental periods:

$$T_{m,n} = 0.006H + 0.164\,(\text{sec}),\ [6 \le H \le 21\,\text{m}] \tag{11}$$

$$T_{R,n} = 0.006H + 0.073\,(\text{sec}),\ [6 \le H \le 21\,\text{m}] \tag{12}$$

$$T_{m,p} = 0.011H + 0.107\,(\text{sec}),\ [9 \le H \le 21\,\text{m}] \tag{13}$$

$$T_{R,p} = 0.011H + 0.048\,(\text{sec}),\ [9 \le H \le 21\,\text{m}] \tag{14}$$

Note that the structural types of RC buildings in Southern Europe differ from those in USA due to the fact that the first usually have stiff infill panels (brick masonry) with an influence in building's total stiffness and hence to its natural period of vibration. Assuming that a two-storey building is considered of type «a» ("low" height), a four-storey of type «b» ("middle" height in Greece) and a seven-storey of type «c» ("high" height in Greece) the fundamental period T_1 has been estimated based on the upper expressions: a) Buildings with "low" height with regular infills: $T_{n,2}$ = 0.11 -

0.20 sec (Equations (11) and (12)). b) Buildings with "middle" height with regular infills: $T_{n,4}$ = 0.15 - 0.24 sec (Equations (11) and (12)). c) Buildings with "middle" height with pilotis: $T_{p,4}$ = 0.18 - 0.24 sec (Equations (13) and (14)). d) Buildings with "high" height with regular infills: $T_{n,7}$ = 0.20 - 0.29 sec (Equations (11) & (12)). e) Buildings with "high" height with pilotis: $T_{p,7}$ = 0.28 - 0.34 sec (Equations (13) and (14)).

According to the EC8 provisions which reflect the design and construction practices of the European building stock the period of vibration has been calculated (Table 1) for: a) Reinforced concrete MRF buildings with "low" height: $T_{MRF,2}$ = 0.29 sec (Equation (3)). b) Reinforced concrete MRF buildings with "middle" height: $T_{MRF,4}$ = 0.48 sec (Equation (3)). c) Reinforced concrete MRF buildings with "high" height: $T_{MRF,7}$ = 0.74 sec (Equation (3)). d) Reinforced concrete SW buildings with "low" height: $T_{SW,2}$ = 0.19 sec (Equation (10)). e) Reinforced concrete SW buildings with "middle" height: $T_{SW,4}$ = 0.32 sec (Equation (10)). f) Reinforced concrete SW buildings with "high" height: $T_{SW,7}$ = 0.49 sec (Equation (10)). Figure 1 presents the stations with the recorded accelerograms in several regions of Athens and the epicentre of the 7th-9-1999 earthquake.

From the notice of the estimated fundamental periods according to the presented formulas for Greek reinforced concrete buildings it comes up the result that the characteristic period T_R (95% reliability) value (0.20) of "low" height RC buildings with regular infills (n-normal) [20] is similar to the estimated value (0.19) according to EC8 for seismic design regarding the "low" height RC shear wall buildings T. The respective value (0.11) of the mean period T_m (50% reliability) is lower, whereas the period is greater (0.29) for "low" rise RC MRF of EC8 design provisions. In addition, both the characteristic (0.15 & 0.18) and mean period (0.24) values are almost the same for "middle" height RC buildings with regular infills (n normal) and with ground levels without infill panels (ppilotis) [20]. As far as the EC8 "middle" rise RC buildings are concerned the period (0.32) is close regarding SW buildings whereas it presents a disparity for MRF. The occasion is also the same regarding "high" rise RC buildings: similar characteristic (0.20 & 0.28) and mean period (0.29 & 0.34) values for RC buildings with regular infills (n-normal) and with ground levels without infill panels (ppilotis) [20], a relatively close period value (0.49) regarding the corresponding EC8 shear wall buildings and a great difference (0.74) regarding the building type of RC moment resisting frames.

Table 1. Estimated fundamental periods.

Number of floors	$T_{1,n}$ (sec) (Karabinis 1986)	$T_{1,p}$ (sec) (Karabinis 1986)	$T_{1,MRF}$ (sec) (EC8)	$T_{1,SW}$ (sec) (EC8)
2	0.11 - 0.20	-	0.29	0.19
4	0.15 - 0.24	0.18 - 0.24	0.48	0.32
7	0.20 - 0.29	0.28 - 0.34	0.74	0.49

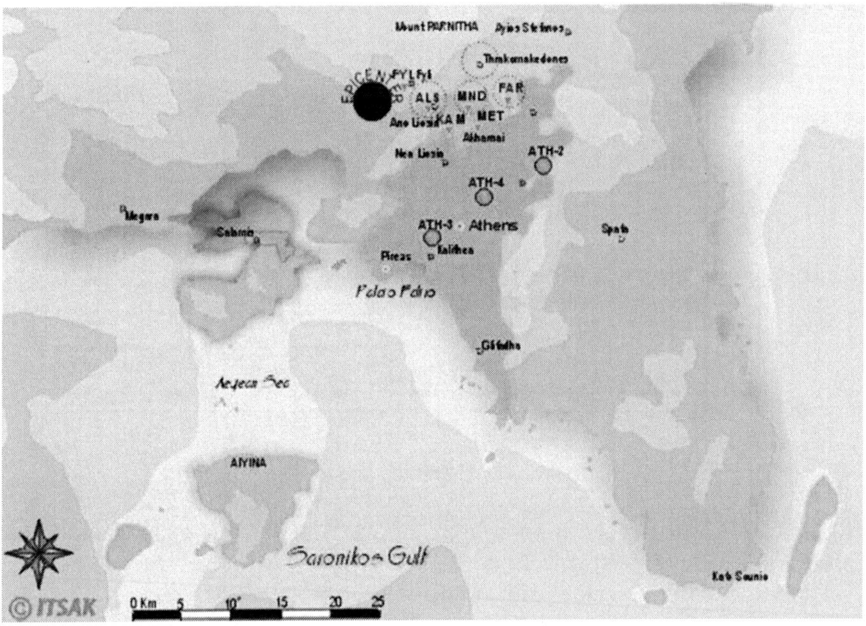

Figure 1. Stations with the recorded accelerograms in Athens and the epicentre of the 7th-9-1999 earthquake.

The elastic acceleration spectra of artificial seismic ground motions and accelerograms are presented in Figures 2 and 3 [21]. In the same figure are also presented the estimated fundamental periods for "low", "medium" and "high" buildings along with the elastic spectrum of the first (1959) and the contemporary (2003) Greek Seismic Code. The percentage of damaged buildings in each height level represents the 29.78% (type «a»), 8.58% (type «b») and 1.65% (type «c») of the total number of buildings (164,135) in the studied area, as it has resulted from the damage data analysis (Table 2).

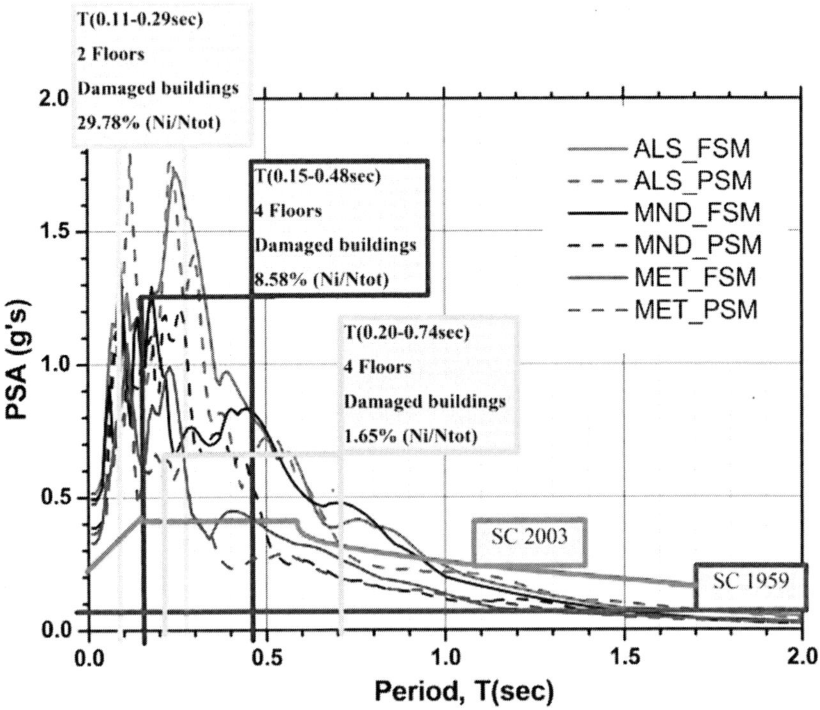

Figure 2. Elastic acceleration spectra of artificial seismic ground motions and the 1st (Greek Seismic Code of 1959) along with the recent design seismic spectrum (Greek Seismic Code of 2003).

3. DAMAGE DATA AFTER THE 7-9-1999 PARNITHA'S EARTHQUAKE

A near field earthquake with moderate to large magnitude, M = 5.9R (according to the Institute of Geodynamics of the National Observatory of Athens) occurred on the 7th of September, 1999 at 14:56 local time (11:56 GMT) with epicentral distance of 18 km from the historical centre of the city of Athens in Greece. The epicenter is located south of the mountain Parnitha, close to the Saronikos Gulf. The meizoseismal area (Figure 4) was considered before this seismic event as of low seismic activity. Although the earthquake magnitude was moderate, the damage was very serious, since the intensity reached IX as it can be observed in existing isoseismal intensity maps of Figure 5 [5-7,22]. Over a hundred of buildings collapsed

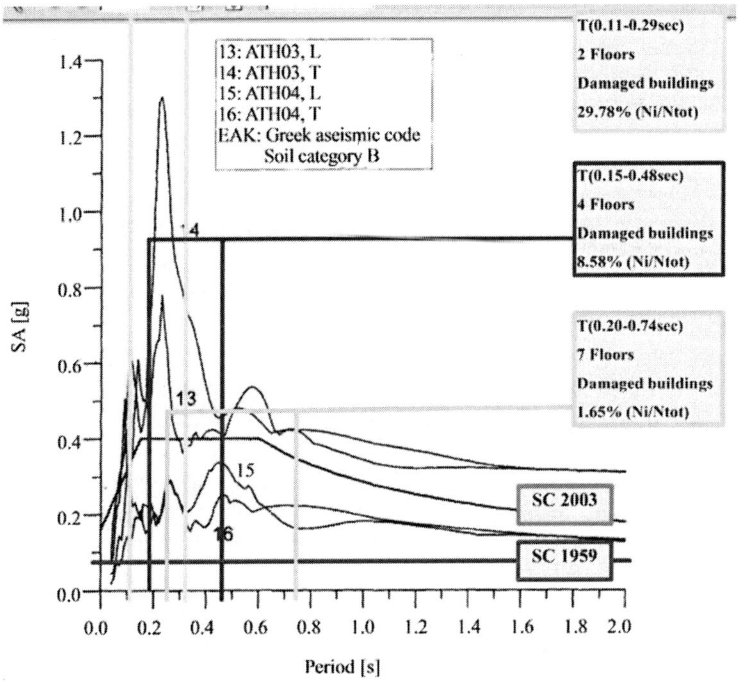

Figure 3. Elastic response spectra of the accelerograms ATHS03 & ATH04 and the 1st (Greek Seismic Code of 1959) along with the recent design seismic spectrum (Greek Seismic Code of 2003).

and thousands sustained considerable damage, causing 143 casualties, about 1000 injuries and thousands of people became homeless. It is the worst natural disaster reported in the modern history of Greece regarding the economic loss. Despite the fact that damage displayed significant differentiation from place to place, the most serious damages were observed at the northern suburbs (e.g. Ano Liosia, Aharnes, Fyli, Menidi, Ilion, Filadelpheia, Kamatero, Zefyri) which was closer to the epicentral area.

The observational database is developed after the first and/or the second round (or level) of inspections, which have been conducted in several regions of Athens, based on instructions provided by the Earthquake Planning and Protection Organization (EPPO) of Greece. The entire initial observational data was collected from different sources (the Post-Earthquake Crisis Management Division of Axarnes region (Filadelfeia-Axarnes-Ano Liosia) and of Piraeus region (Piraeus-down town of Athens-

PeristeriEleusina) and the National Service for the Rehabilitation of Earthquake Victims) and raised the enormous number of 535,152 reports of post-earthquake surveys. After filtering the collected data (checking one-by-one) and eliminating duplicate reports a unified total database has been created referring to the extended urban region of Attica. This dataset consists of 296,919 unique inspections having avoided the overestimation of the damage level. It is essential to clarify that the number of the above mentioned inscriptions refers to the number of autopsies and does not coincide with the number of buildings. A new process for the unified database has been followed, resulting that the 296,919 inspections are associated with 180,945 damaged buildings. It is noted that many of the 180,945 buildings were not fully described and the corresponding buildings have been disregarded from the process. Information about the total number of buildings per structural type for the regions mentioned in the database is also provided by the National Statistics Agency of Greece (N.S.S.G.) according to the results of the 2000-1 census. Comparing the total number of damaged buildings (180,945) to the total number of buildings in the affected area (753,078) it is concluded that the dataset addresses to 24.03% of the total local population of buildings, which is a wide and reliable statistical sample [5-7, 22].

Table 2. Correlation of the estimated fundamental periods with the level of the recorded damage and the number of floors for 164,135 buildings.

Number of Floors	T_1 (sec) (Karabinis 1986)	T_1 (sec) (EC8)	Damage Level				Total Number of Buildings
			Light (Green)	Moderate (Yellow)	Extensive (Red)	Collapse (Black)	
1	$T_{1,R} = 0.11$ $T_{1,m} = 0.20$ (regular infills)	0.19 (SW) 0.29 (MRF)	(22.496)	(16.913)	(2778)	(1208)	(43.395)
			13.71%	10.30%	1.69%	0.74%	26.44% (N_i/N_{tot})
2			(28.952)	(17.723)	(1567)	(644)	(48.886)
			17.64%	10.80%	0.95%	0.39%	29.78% (N_i/N_{tot})
3			(21.230)	(12.014)	(728)	(266)	(34.238)
			12.93%	7.32%	0.44%	0.16%	20.86% (N_i/N_{tot})
4	$T_{1,R} = 0.15$ $T_{1,m} = 0.24$ (regular infills)	0.32 (SW) 0.48 (MRF)	(10.084)	(3795)	(158)	(48)	(14.085)
			6.14%	2.31%	0.10%	0.03%	8.58% (N_i/N_{tot})
5	$T_{1,R} = 0.18$ $T_{1,m} = 0.24$ ("pilotis")		(9315)	(1826)	(47)	(18)	(11.206)
			5.68%	1.11%	0.03%	0.01%	6.83% (N_i/N_{tot})
6			(7120)	(1015)	(20)	(2)	(8157)
			4.34%	0.62%	0.01%	0.00%	4.97% (N_i/N_{tot})
7	$T_{1,R} = 0.20$ $T_{1,m} = 0.29$ (regular infills)	0.49 (SW) 0.74 (MRF)	(2270)	(427)	(9)	(0)	(2706)
			1.38%	0.26%	0.01%	0.00%	1.65% (N_i/N_{tot})
≥8	$T_{1,R} = 0.28$ $T_{1,m} = 0.34$ ("pilotis")		(1187)	(256)	(17)	(2)	(1462)
			0.72%	0.16%	0.01%	0.00%	0.89% (N_i/N_{tot})
Total number of buildings			(102.694)	(53.969)	(5324)	(2188)	(164.135) (N_{tot})
			62.54%	32.88%	3.24%	1.33%	100.00%

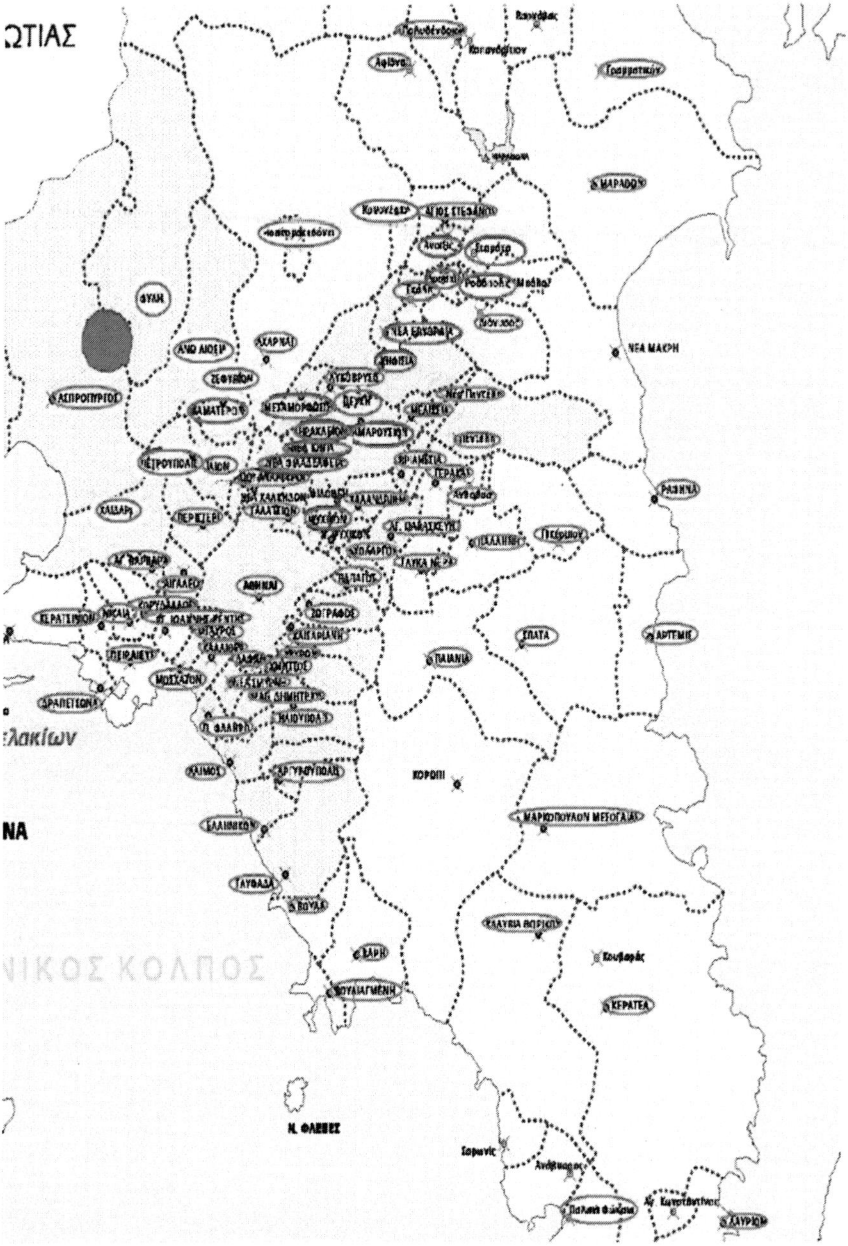
ΦΥΛΗ
ΑΣΠΡΟΠΥΡΓΟΣ
ΑΝΩ ΛΙΟΣΙΑ
ΑΧΑΡΝΑΙ
ΖΕΦΥΡΙΟΝ
ΠΕΤΡΟΥΠΟΛΙΣ
ΧΑΙΔΑΡΙ
ΠΕΡΙΣΤΕΡΙ
ΑΙΓΑΛΕΩ
ΑΘΗΝΑΙ
ΚΕΡΑΤΣΙΝΙΟΝ
ΠΕΙΡΑΙΕΥΣ
ΜΟΣΧΑΤΟΝ
ΔΡΑΠΕΤΣΩΝΑ
ΖΩΓΡΑΦΟΣ
ΑΛΙΜΟΣ
ΕΛΛΗΝΙΚΟΝ
ΓΛΥΦΑΔΑ
ΒΟΥΛΑ
ΒΑΡΗ
ΒΟΥΛΙΑΓΜΕΝΗ
ΚΗΦΙΣΙΑ
ΜΑΡΑΘΩΝ
ΝΕΑ ΜΑΚΡΗ
ΡΑΦΗΝΑ
ΠΑΙΑΝΙΑ
ΣΠΑΤΑ
ΑΡΤΕΜΙΣ
ΚΟΡΩΠΙ
ΜΑΡΚΟΠΟΥΛΟΝ ΜΕΣΟΓΑΙΑΣ
ΚΕΡΑΤΕΑ
ΛΑΥΡΙΟΝ
Ν. ΦΛΕΒΕΣ
ΝΑ
ΝΙΚΟΣ ΚΟΛΠΟΣ

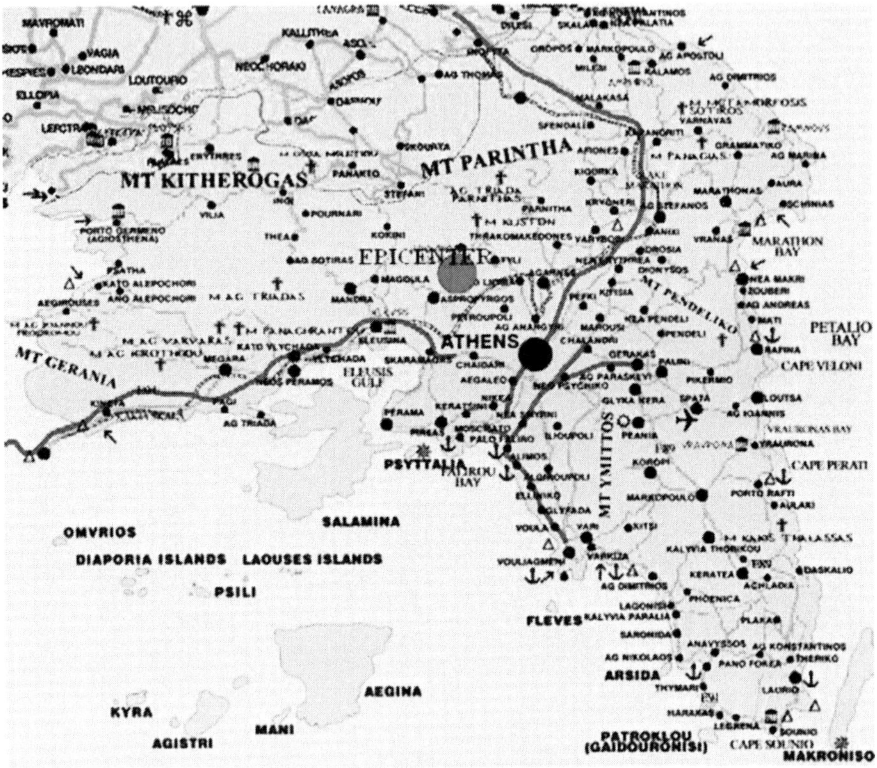

Figure 4. Meizoseimal area of Attica with the regions of the observational damage data.

The vulnerability assessment requires a classification system to characterize the earthquake-exposed building stock and correlate it with the developed damage. This needs the division of buildings into groups with similar seismic behaviour due to a probable earthquake. In the current research, apart from the characteristics that affect the seismic response of a structure, the proposed classification system is also dependent on the information collected from the post earthquake surveys. Unfortunately, the existence or not of ground levels without infill panels (pilotis) or other irregularities, which may influence the development of earthquake damage, is not known. In the statistical database, the structural systems are divided

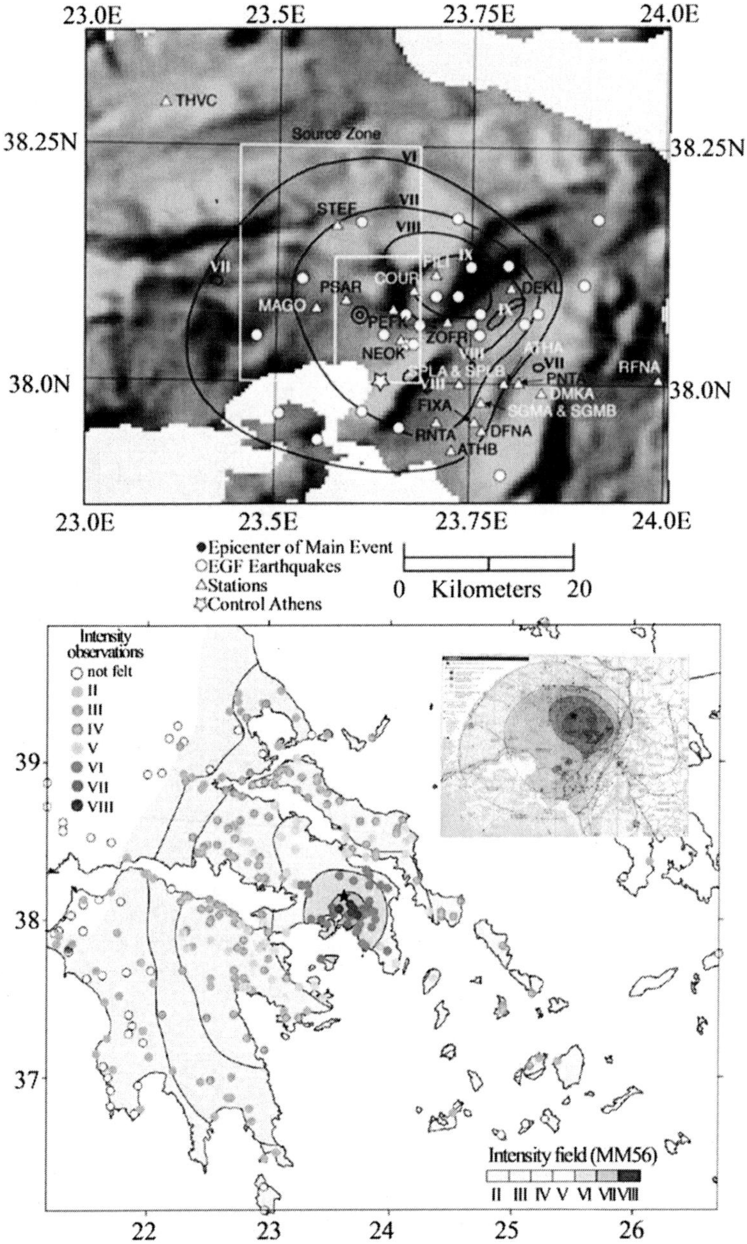

Figure 5. Existing isoseismal intensity maps [23,24].

into four groups: 1) Reinforced concrete buildings (RC) with moment resisting frames or dual system (frame + shear walls); 2) Mixed buildings (MIX) with vertical bearing structure constituted by elements of both masonry and reinforced concrete; 3) Masonry buildings (MAS) with vertical elements of masonry and horizontal elements of reinforced concrete, metal or wood; and 4) Other buildings (OTH), which typically include any buildings not belonging to the previous groups. The reinforced concrete structures are further classified based on the different seismic code periods at the time of their design: RC1: without a seismic code or during the period 1959-1985; RC2: during the period 1985-1995; and RC3 after 1995. The mixed structures are further classified into MIX1, MIX2, and MIX3 using identical criteria. The threshold of each period is identified with a change in Greek Seismic Codes. The level of seismic design and detailing for R/C buildings in Greece, could generally be discriminated in four subclasses, as follows: 1) No Seismic Code (before 1959): with very low level or no seismic design and poor quality of detailing; 2) Low Seismic Code (1959-1985: the 1st Seismic Code of 1959 corresponding approximately to pre-1980s codes in Southern Europe) with low level of seismic design; 3) Moderate Seismic Code (1985-1995: the 1st Seismic Code of 1959 with the 1985 Supplement Clauses corresponding approximately to post-1980s codes in Southern Europe): with medium level of seismic design and reasonable seismic detailing; 4) High Seismic Code (after 1995: New Greek Seismic Code of 2003, similar to Eurocode 8): with adequate level of seismic design according to the new generation of seismic codes and ductile seismic detailing of R/C members. Buildings constructed before and after the introduction of the first Seismic Code are often treated similarly in Greece according to the National Technical Chamber of Greece [25]. Even if many buildings have been disregarded from the database due to the lack of information, such as the structural type or the date of construction, the presented here building stock still remains a wide statistical database.

4. SEISMIC DAMAGE AND FUNDAMENTAL PERIOD

After the time-consuming statistical elaboration and analysis of the database, the correlation for 164,135 buildings of the degree of damage with the number of floors is presented in Table 2 and in Figure 6. The estimated fundamental periods for existing reinforced concrete buildings of several heights ("low", "medium" and "high") and structural types (moment resisting frames-MRF, shearwall-SW, with regular infills or no infills on the ground floor-pilotis) are correlated to the different levels of the recorded damage.

The information from the database refers to qualitative characterizations of damage level used in the post-earthquake surveys in Greece, based on instructions provided by EPPO in 1984 and in 1997 in order to define whether its seismic capacity is adequate against future expected seismic forces. The damage level characterized by the inspection crews, as follows: a) Green: building with no or light damage, or building whose earthquake resistance has not been reduced; b) Yellow: building with moderate damage and reduced earthquake resistance; c) Red: building with very heavy damage or partial collapse; and d) Collapse: building that has collapsed or is under demolition. A damage scale for the measurable recording, beyond the pre-mentioned qualitative characterization Correlation for 164,135 buildings of the degree of damage with the number of floors of seismic damage in Greek post-earthquake surveys, is presented in this research. Furthermore, the qualitative characterization of damage has been correlated to the different damage states of the scale for the calibration of seismic damage referring to the main structural types of RC buildings [26-28].

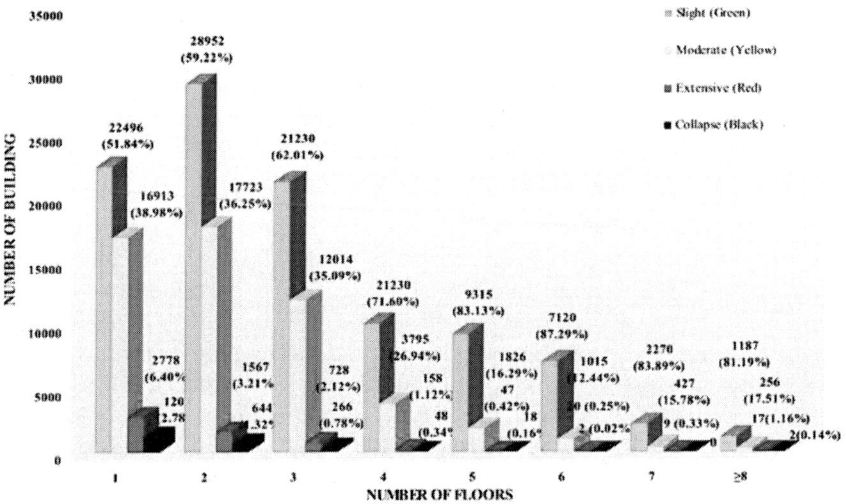

Figure 6. Correlation of the estimated fundamental periods with the level of the recorded damage and the number of floors for 164,135 buildings.

The correlation for 30149 RC1 buildings of the damage level with the number of floors is presented in Table 3 and in Figure 7. Table 4 and Figure 8 present the correlation of the damage level with the number of floors for 4501 RC2 buildings. The correlation for 1591 RC3 buildings of the damage level with the number of floors is presented in Table 5 and in Figure 9. The estimated fundamental periods for existing RC buildings of several heights are also correlated to the different levels of the recorded damage. The correlation of the damage level buildings with the number of floors for 15654 MIX1 is presented in Table 6 and in Figure 10. Table 7 and Figure 11 present the correlation of the damage level with the number of floors for 538 MIX2 buildings. Finally, the correlation for 87 MIX3 buildings of the damage level with the number of floors is presented in Table 8 and in Figure 12.

5. SEISMIC DAMAGE SCALE

The seismic behavior of the structures during an earthquake represents an experiment in a physical scale (1:1) and constitutes the most objective examination of the sufficiency of seismic codes and construction techniques. However, the reliable estimation of buildings response depends on the used method for the recording of seismic damage. The latter may vary in detail (approximate, analytical, etc.) and extent (numerous buildings, group of buildings or a single structure).

The structural damage can be considered the greatest cause of life and monetary loss in most earthquakes and can be used to assess the performance of buildings. In each database, this is described in terms of either damage state or percentage of loss. Due to the different parameters that influence the recording of damage (building types, seismic design codes, performance levels, etc.), it cannot be easily compared nor is it easy to combine and compare damage data. Comparisons of the recording damage have been presented in the literature [5,26,28-31]. The description of damage of every performance level is often mainly based on ATC-13 [32]. This is also the case for the damage calibration regarding the economic damage index in Greece [25,33] and in several other vulnerability studies [34,35]. Table 9 has been created in order to be able to implement a correlation analysis [36]. It correlates the most important existing damage scales, with the existing method used for the recording

Table 3. Correlation of the estimated fundamental period T_1 with the level of the recorded damage for 30,149 buildings of RC1 structural type with low level of seismic design.

Number of Floors	T_1 (sec) (Karabinis 1986)	T_1 (sec) (EC8)	Damage Level for RC1 Structural Type (moment resisting frames or frame-wall earlier than 1985)				
			Light (Green)	Moderate (Yellow)	Extensive (Red)	Collapse (Black)	Total Number of Buildings
1	$T_{1,R} = 0.11$ $T_{1,m} = 0.20$ (regular infills)	0.19 (SW) 0.29 (MRF)	(1017)	(1948)	(110)	(52)	(3127)
			3.37%	6.46%	0.36%	0.17%	10.37% (N_i/N_{tot})
2			(2699)	(5771)	(252)	(133)	(8855)
			8.95%	19.14%	0.84%	0.44%	29.37% (N_i/N_{tot})
3			(2745)	(6248)	(227)	(94)	(9314)
			9.10%	20.73%	0.75%	0.31%	30.89% (N_i/N_{tot})
4	$T_{1,R} = 0.15$ $T_{1,m} = 0.24$ (regular infills)	0.32 (SW) 0.48 (MRF)	(1450)	(2380)	(59)	(15)	(3904)
			4.81%	7.89%	0.20%	0.05%	12.95% (N_i/N_{tot})
5	$T_{1,R} = 0.18$ $T_{1,m} = 0.24$ ("pilotis")		(1044)	(1196)	(15)	(7)	(2262)
			3.46%	3.97%	0.05%	0.02%	7.50% (N_i/N_{tot})
6			(910)	(718)	(10)	(1)	(1639)
			3.02%	2.38%	0.03%	0.00%	5.44% (N_i/N_{tot})
7	$T_{1,R} = 0.20$ $T_{1,m} = 0.29$ (regular infills)	0.49 (SW) 0.74 (MRF)	(356)	(309)	(3)	(0)	(668)
			1.18%	1.02%	0.01%	0.00%	2.22% (N_i/N_{tot})
≥8	$T_{1,R} = 0.28$ $T_{1,m} = 0.34$ ("pilotis")		(186)	(190)	(3)	(1)	(380)
			0.62%	0.63%	0.01%	0.00%	1.26% (N_i/N_{tot})
Total number of buildings			(10,407)	(18,760)	(679)	(303)	(30,149) (N_{tot})
			34.52%	62.22%	2.25%	1.01%	100.00%

Correlation for 30,149 RC1 buildings of the degree of damage with the number of floors

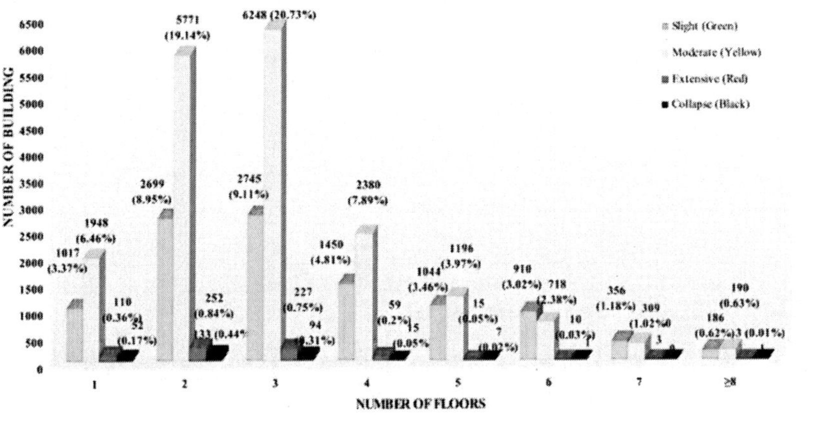

Figure 7. Correlation of the estimated fundamental period T_1 with the level of the recorded damage for 30,149 buildings of RC1 structural type with low level of seismic design.

Table 4. Correlation of the estimated fundamental period T₁ with the level of the recorded damage for 4501 buildings of RC2 structural type with medium level of seismic design.

Number of Floors	T_1 (sec) (Karabinis 1986)	T_1 (sec) (EC8)	Damage Level for RC2 Structural Type (moment resisting frames or frame-wall between 1985-1995) Light (Green)	Moderate (Yellow)	Extensive (Red)	Collapse (Black)	Total number of buildings
1			(146)	(256)	(5)	(6)	(413)
	$T_{1,R}$ = 0.11 $T_{1,m}$ = 0.20 (regular infills)	0.19 (SW) 0.29 (MRF)	3.24%	5.69%	0.11%	0.13%	9.18% (N_i/N_{tot})
2			(523)	(1078)	(43)	(30)	(1674)
			11.62%	23.95%	0.95%	0.67%	37.19% (N_i/N_{tot})
3			(408)	(800)	(37)	(12)	(1257)
			9.06%	17.77%	0.82%	0.27%	27.93% (N_i/N_{tot})
4	$T_{1,R}$ = 0.15 $T_{1,m}$ = 0.24 (regular infills)		(203)	(305)	(6)	(1)	(515)
		0.32 (SW) 0.48 (MRF)	4.51%	6.78%	0.13%	0.02%	11.44% (N_i/N_{tot})
5	$T_{1,R}$ = 0.18 $T_{1,m}$ = 0.24 ("pilotis")		(181)	(210)	(5)	(2)	(398)
			4.02%	4.67%	0.11%	0.04%	8.84% (N_i/N_{tot})
6			(88)	(67)	(2)	(0)	(157)
			1.96%	1.49%	0.04%	0.00%	3.49% (N_i/N_{tot})
7	$T_{1,R}$ = 0.20 $T_{1,m}$ = 0.29 (regular infills)		(22)	(28)	(1)	(0)	(51)
		0.49 (SW) 0.74 (MRF)	0.49%	0.62%	0.02%	0.00%	1.13% (N_i/N_{tot})
≥8	$T_{1,R}$ = 0.28 $T_{1,m}$ = 0.34 ("pilotis")		(26)	(9)	(1)	(0)	(36)
			0.58%	0.20%	0.02%	0.00%	0.80% (N_i/N_{tot})
Total number of buildings			(1597)	(2753)	(100)	(51)	4501 (N_{tot})
			35.24%	61.16%	2.22%	1.13%	100.00%

Correlation for 4.501 RC2 buildings of the degree of damage with the number of floors

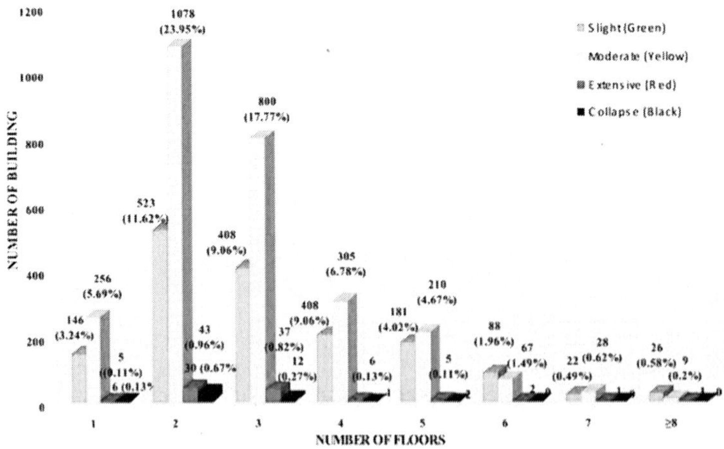

Figure 8. Correlation of the estimated fundamental period T₁ with the level of the recorded damage for 4501 buildings of RC2 structural type with medium level of seismic design.

Table 5. Correlation of the estimated fundamental periods with the level of the recorded damage for 1591 buildings of RC3 structural type with high level of seismic design.

Number of Floors	T_1 (sec) (Karabinis 1986)	T_1 (sec) (EC8)	Damage Level for RC3 Structural Type (moment resisting frames or frame-wall after 1995)				
			Light (Green)	Moderate (Yellow)	Extensive (Red)	Collapse (Black)	Total Number of Buildings
1	$T_{1,R}$ = 0.11 $T_{1,m}$ = 0.20 (regular infills)	0.19 (SW) 0.29 (MRF)	(57)	(101)	(3)	(2)	(163)
			3.58%	6.35%	0.19%	0.13%	10.25% (N_i/N_{tot})
2			(194)	(364)	(10)	(9)	(577)
			12.19%	22.88%	0.63%	0.57%	36.27% (N_i/N_{tot})
3			(127)	(206)	(15)	(3)	(351)
			7.98%	12.95%	0.94%	0.19%	22.06% (N_i/N_{tot})
4	$T_{1,R}$ = 0.15 $T_{1,m}$ = 0.24 (regular infills)	0.32 (SW) 0.48 (MRF)	(66)	(69)	(3)	(2)	(140)
			4.15%	4.34%	0.19%	0.13%	8.80% (N_i/N_{tot})
5	$T_{1,R}$ = 0.18 $T_{1,m}$ = 0.24 ("pilotis")		(99)	(81)	(5)	(0)	(185)
			6.22%	5.09%	0.31%	0.00%	11.63% (N_i/N_{tot})
6			(57)	(47)	(0)	(0)	(104)
			3.58%	2.95%	0.00%	0.00%	6.54% (N_i/N_{tot})
7	$T_{1,R}$ = 0.20 $T_{1,m}$ = 0.29 (regular infills)	0.49 (SW) 0.74 (MRF)	(25)	(27)	(0)	(0)	(52)
			1.57%	1.70%	0.00%	0.00%	3.27% (N_i/N_{tot})
≥8	$T_{1,R}$ = 0.28 $T_{1,m}$ = 0.34 ("pilotis")		(14)	(5)	(0)	(0)	(19)
			0.88%	0.31%	0.00%	0.00%	1.19% (N_i/N_{tot})
Total number of buildings			(639)	(900)	(36)	(16)	(1591)
			40.16%	56.57%	2.26%	1.01%	100.00%

Correlation for 1,591 RC3 buildings of the degree of damage with the number of floors

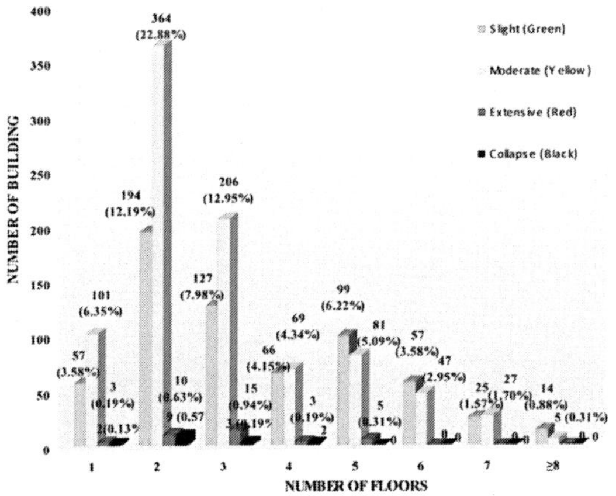

Figure 9. Correlation of the estimated fundamental periods with the level of the recorded damage for 1591 buildings of RC3 structural type with high level of seismic design.

Table 6. Correlation of the level of the recorded damage with the number of floors for 15,654 buildings of MIX1 structural type with low level of seismic design.

Number of Floors	Damage State for MIX1 Structural Type (mixed buildings with vertical bearing structure by elements of both masonry and RC earlier than 1985)				
	Slight (Green)	Moderate (Yellow)	Extensive (Red)	Collapse (Black)	
1	1145	3155	235	140	
	7.31%	20.15%	1.5%	0.89%	
2	1687	4635	268	141	
	10.77%	29.61%	1.71%	0.90%	
3	770	2228	111	60	
	4.92%	14.23%	0.71%	0.38%	
4	188	516	15	14	
	1.20%	3.30%	0.10%	0.09%	
5	49	129	1	2	
	0.31%	0.82%	0.01%	0.01%	
6	40	45	2	0	
	0.26%	0.29%	0.01%	0.00%	
7	16	23	0	0	
	0.10%	0.15%	0.00%	0.00%	
≥8	11	22	5	1	
	0.07%	0.14%	0.03%	0.01%	
Total	3906	10.753	637	358	15.654
	24.95%	68.69%	4.07%	2.29%	100.00%

Correlation for 164,135 buildings of the degree of damage with the number of floors

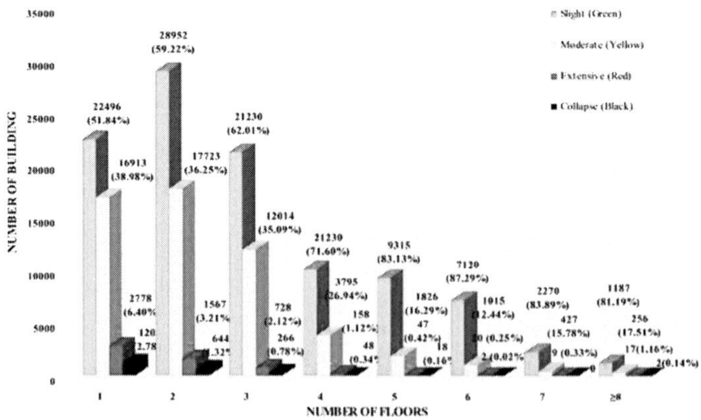

Figure 10. Correlation of the level of the recorded damage with the number of floors for 15,654 buildings of MIX1 structural type with low level of seismic design.

Table 7. Correlation of the level of the recorded damage with the number of floors for 538 buildings of MIX2 structural type with medium level of seismic design.

Number of Floors	Damage State for MIX2 Structural Type (mixed buildings with vertical bearing structure by elements of both masonry and RC 1985-1995)				
	Slight (Green)	Moderate (Yellow)	Extensive (Red)	Collapse (Black)	
1	37	88	13	3	
	6.88%	16.36%	2.42%	0.56%	
2	45	111	42	3	
	8.36%	20.63%	7.81%	0.56%	
3	24	52	29	4	
	4.46%	9.67%	5.39%	0.74%	
4	8	10	25	1	
	1.49%	1.86%	4.65%	0.19%	
5	7	8	9	2	
	1.30%	1.49%	1.67%	0.37%	
6	1	5	2	0	
	0.19%	0.93%	0.37%	0.00%	
7	0	1	3	0	
	0.00%	0.19%	0.56%	0.00%	
≥8	1	0	4	0	
	0.19%	0.00%	0.76%	0.00%	
Total	123	275	127	13	538
	22.86%	51.12%	23.61%	2.42%	100.00%

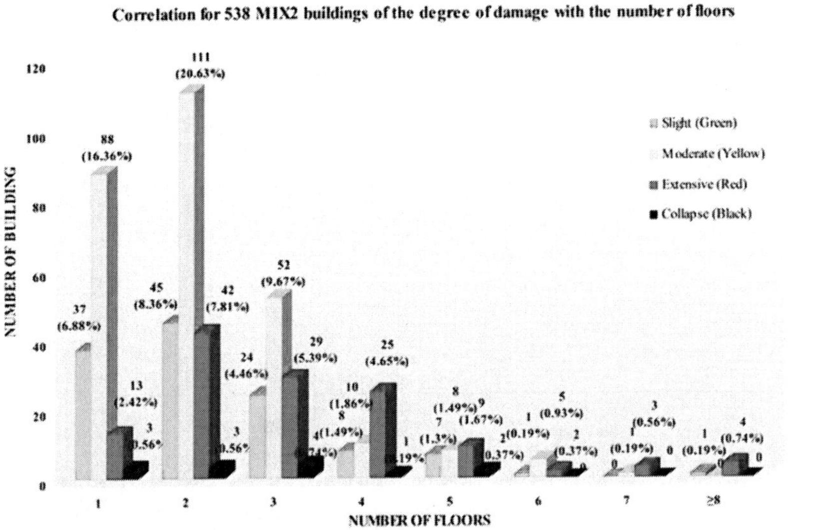

Figure 11. Correlation of the level of the recorded damage with the number of floors for 538 buildings of MIX2 structural type with medium level of seismic design.

Table 8. Correlation of the level of the recorded damage with the number of floors for 87 buildings of MIX3 structural type with high level of seismic design.

Number of Floors	Damage State for MIX3 Structural Type (mixed buildings with vertical bearing structure by elements of both masonry and RC after 1995)				
	Slight (Green)	Moderate (Yellow)	Extensive (Red)	Collapse (Black)	
1	9 10.47%	18 20.93%	2 2.33%	0 0.00%	
2	11 12.79%	18 20.93%	5 5.81%	1 1.16%	
3	4 4.65%	10 11.63%	1 0.00%	0 0.00%	
4	1 1.16%	3 3.49%	0 0.00%	0 0.00%	
5	2 2.33%	0 0.00%	0 0.00%	0 0.00%	
6	0 0.00%	0 0.00%	0 0.00%	0 0.00%	
7	0 0.00%	1 1.16%	0 0.00%	0 0.00%	
≥8	0 0.00%	1 1.16%	0 0.00%	0 0.00%	
Total	27 31.40%	51 59.30%	7 8.14%	1 1.16%	87 100.00%

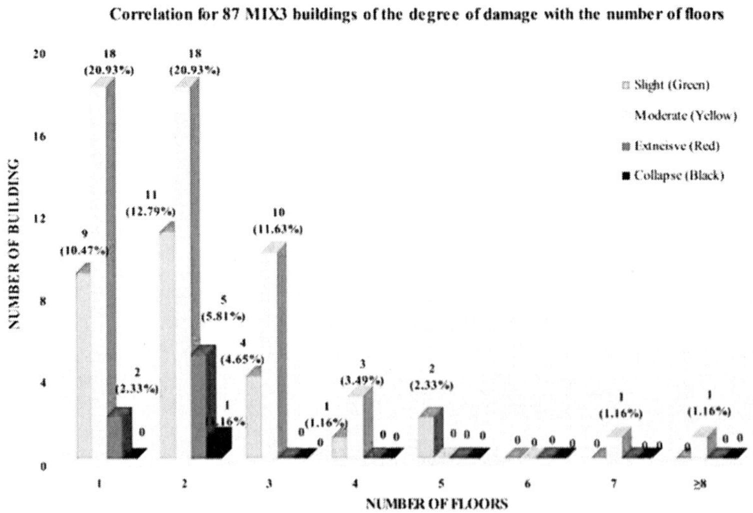

Figure 12. Correlation of the level of the recorded damage with the number of floors for 87 buildings of MIX3 structural type with high level of seismic design.

and classification of damage in Greece and the presented in this research damage scale.

The forms of post-earthquake surveys in Greece commonly use the method of Rapid Visual Screening (R.V.S.), which is based on a macroscopic inspection of the building in order to define whether its seismic capacity is adequate against future expected seismic forces. The characterization of damage by inspection crews in the 7-9- 1999 Athens earthquake was based on instructions provided by EPPO 1997 [37] and to a lesser degree on the EPPO instructions of 1984 [38]. In addition, the vulnerability models proposed by National Technical Chamber of Greece [25,39] were mostly based on a hybrid methodology involving elements from both empirically and analytically calculated structural damage indices which have been correlated to monetary loss [21,33,40].

The observational database is developed after the first or/and the second round (or level) of inspections, which have been conducted in several regions of Athens, based on instructions provided by Earthquake Planning and Protection Organization (EPPO) of Greece. The first round of inspection is a rapid visual screening method conducted by a couple team of engineers in order to define in a short period of time the seismic safety of numerous structures. A second and more detailed level of inspection is followed for those buildings with inadequate estimated seismic performance in order to prioritize them for further more detailed analytical assessment that it would be required to design a rehabilitation scheme. However, the information from the current mentioned database refers only to quailtative characterizations of damage level by the inspection crews as it has been already mentioned: a) Green: building with no or light damage, or building whose earthquake resistance has not been reduced; b) Yellow: building with moderate damage and reduced earthquake resistance; c) Red: building with very heavy damage or partial collapse; and d) Collapse: building that has collapsed or is under demolition. In the collected data, there was no information about the cost of repairs or the description of damage. The need for a measurable calibration of the recorded damage is obvious.

The distribution of buildings (180,427) in the created database according to the pre-described degree of damage is presented in Table 10. Specifically: a) the 2716 structures which were characterised as under demolitioncollapse (black) represent the 1.51% of the total damaged population of buildings; b) the 6423 structures with extensive non-repairable damages (red) to the structural system represent the 3.56%; c) the 56,533 structures with moderate (yellow) repairable damages represent the 31.33% of the total damaged buildings in Attica; and d) most of the structures (114,755) with light (green) repairable damages represent the 63.60%, respectively.

A damage scale is presented for the reinforced concrete (RC) buildings wherein a calibration of seismic damage is presented beyond the qualitative description of the performance levels (Table 11). The use of the new scale does not require specialized knowledge, and is intentionally quite similar to the familiar to many engineers, manner of estimation of post-earthquake damage. The scale is approximate but rational and can be easily and rapidly applied. The novelty of the new scale is that it introduces an empirical measurable damage index, which depends on the severity and the extent of damage, right from the field where in situ inspection is conducted. This characteristic is lacking from the existing scales of measurement. Its inclusion may decrease the need for second round timeconsuming sampling inspections of damaged buildings [26-28].

Table 9. Correspondence between damage states definitions and damage factors ranges (%).

Vision 2000 (SEAOC 1995)		FEMA 273		HAZUS99		EMS98		ATC13		EPPO (2001 & 2006)		PROPOSED EPPO	
None (0)													
Fully Operational	0 - 2	Operational	0 - 1	None	0 - 2	Grade 1	0 - 1	Slight	0 - 1	Slight	0 - 1	Green	0 - 1
Operational	2 - 10	Immediate Ocupancy	1 - 10	Slight	2 - 10	Grade 2	1 - 11	Light	1 - 10	Light	1 - 10	Yellow	1 - 30
								Moderate	10 - 30	Moderate	10 - 30		
Life Safe	10 - 50	Life Safety	10 - 30	Moderate	10 - 50	Grade 3	11 - 50						
								Heavy	30 - 60	Heavy	30 - 60		
Near Collapse	50 - 100	Collapse Prevention	30 - 100	Extensive	50 - 100	Grade 4	50 - 100	Major	60 - 100	Major	60 - 100	Red	30 - 100
Collapse (100)													

Table 10. Distribution of buildings according to the degree of damage (180,427 buildings).

Damage State	Number of Buildings (N_i)	N_i/N_{tot} (%)	N_i/75,3078 (%)
Slight (Green)	114.755 (N_1)	63.60%	15.24%
Moderate (Yellow)	56.533 (N_2)	31.33%	7.50%
Extensive (Red)	6423 (N_3)	3.56%	0.85%
Collapse (Black)	2716 (N_4)	1.51%	0.36%
Total	180,427 (N_{tot})	100.00%	23.96%

Table 11. The presented damage scale.

PROPOSED DAMAGE SCALE				TYPICAL STRUCTURAL TYPES OF RC BUILDINGS							
				DUCTILE MRF		NON-DUCTILE MRF		INFILLED MRF		MIXED	
ECONOMIC DAMAGE INDEX	DAMAGE LEVEL	GREECE	FEMA	STRUCTURAL DAMAGE INDEX	DESCRIPTION	STRUCTURAL DAMAGE INDEX	DESCRIPTION	STRUCTURAL DAMAGE INDEX	DESCRIPTION	STRUCTURA DAMAGE INDEX	DESCRIPTION
0%	None	None	None	0.0%	None	0.0%	None	0.00%	None	0.00%	None
0% - 1%	Slight	Green 114735 (N1) 63.60% (N1/180945(%)) 15.24% (N1/753078(%))	Operational	0.0% - 0.2%	Minor cracking in partitions, infills and ceilings, hairline craks of structural elements	0.0% - 0.1%	Minor cracking in partitions, infills and ceilings, hairline craks of structural elements	0.0% - 0.1%	Minor cracking in partitions, infills and ceilings, hairline craks of structural elements	0.0% - 0.2%	Minor cracking in partitions, infills and ceilings, hairline craks of structural elements
1% - 10%	Light		Immediate Occupancy	0.2% - 0.4%	Minor cracks in structural elements, facades and partitions, hairline cracking in beams/columns near joints (<1 mm)	0.1% - 0.2%	Minor cracks in structural elements, facades and partitions, hairline cracking in beams/columns near joints (<1 mm)	0.1% - 0.2%	Cracks at wall-frame interfaces, diagonal cracking of walls, limited crushing of bricks at beam/column connections, star cracking at corners of openings	0.2% - 0.4%	Hairline cracking on shear-walls and coupling beams, onset of concrete spalling
10% - 30%	Moderate	Yellow 56531 (N2) 31.32% (N2/180945(%)) 7.50% (N2/753078(%))	Life Safety	0.4% - 1.0%	Cracks in most beams and columns with larger flexural cracks, yielding in a limited number of structural elements, start of concrete spalling	0.2% - 0.5%	Flexural and shear cracks in most beams and columns, yielding in a limited number of stuctural elements, limited shear cracking and concrete spalling	0.2% - 0.4%	Increased brick crushing at beam-column connections, start of stuctural damage, some diagonal shear cracking in structural elements mostly of the exterior frames	0.4% - 0.9%	Cracks in most shear walls, some walls reach yield capacity, increased diagonal cracking and concrete spalling at wall corners and around openings, extensive cracks in coupling beams
30% - 60%	Extensive	Red 6423 (N3) 3.56% (N3/180945(%)) 0.85% (N3/753078(%))	Collapse Prevention	1.0% - 1.80%	Some structural elements have reached ultimate capacity, extensive flexural cracking, concrete spalling, bar buckling, short column failure, severe joint damage, permanent drift	0.5% - 0.8%	Bar pull-out, loss of bond at lap-splices, broken ties, possible bar buckling and shear failure of structural elements, permanent drift	>0.40%	Extensive cracking of infills falling bricks, partial failure of many infills, out-of-plane bulging, heavier damage in frame members, some shear failure, permanent drift	0.9% - 1.55%	Most shear walls have exceeded yield and some walls reach ultimate capacity, boundary element distress, bar buckling, extensive through-wall cracks, shear failure of some frame members, sliding at joints
60% - 100%	Partial Collapse			1.8% - 3.0%	Collapse of a few columns, a building wing or single upper floor	0.8% - 1.0%	Shear failure of many columns or impending soft-storey failure	0.4% - 0.8%	Shear failure of beams and/or columns causing partial collapse, near total infill failure	1.65% - 2.75%	Coupling beams-panels shattered and virtually disintegrated, some shear walls fail
100%	Collapse	Black 2716 (N4) 1.51% (N4/180945(%)) 0.36% (N4/753078(%))	Collapse	>3.00%	Complete or impending building collapse	>1.00%	Complete or soft-storey failure at ground floor	>0.80%	Complete or impending building collapse	>2.75%	Complete or impending building collapse

The proposed scale is subdivided into seven damage levels, each of which is defined in terms of structural and non-structural damage, which would be expected in a future seismic scenario in the four main structural types of RC buildings that are met in Europe: ductile frames, nonductile frames, frames with masonry walls and mixed buildings. Definition of performance levels in descriptive terms is not sufficient for the development of vulnerability curves. In order to consider the different damage rates of lateral-load systems and hence relate the curves to the building type, the scale must be calibrated to a measurable structural response parameter. In the proposed damage scale the performance levels, ranging from "none" damage to "collapse", are defined according to the physical description of the seismic damage and, as well, in terms of structural and economic damage index. The different drift thresholds, which vary significantly for ductile and nonductile systems, for the associated damage states of each structural type have been adopted by experimental data and theoretical analyses [41,42]. The economic damage index (in monetary loss) expresses the cost of repair as a fraction of the total cost of the building. The calibration of the earthquake damage by presenting both structural and economic damage index in the same scale, allows their direct comparison and correlation. The description of damage in every performance level is based on the main existing damage scales. The proposed calibration for the different levels of seismic damage severity regarding the economic damage index is in accordance with the familiar manner of damage classification in Greece [25]. It must be noted that the proposed methodology regarding the postearthquake surveys and the pre-defined limits of damage severity expressed in monetary losses, is similar to the instructions provided by EPPO and FEMA. Besides, most of the methods for the recording of seismic damage have been adopted by FEMA and ATC. The comparison of the proposed damage scale with FEMA and EPPO is also presented in Table 11. A numerical example is presented in the scale according to the analysis results for the damage distribution of the created database according to the EPPO characterization. The definition of damage levels of the proposed scale and the performance levels of FEMA along with the EPPO results of the database are presented and compared on a typical performance curve for RC frame buildings in Figure 13. In addition, a correlation is fulfilled between the different damage states and the fundamental period regarding building types of RC frames. The numerical values of the drift thresholds in different damage levels for other structural types are given in Table 11. In Figure 13 the damage levels of the proposed scale in terms of structural damage index on the performance curve could be correlated to the damage levels in terms of loss damage index on a qualitative here "primary vulnerability curve".

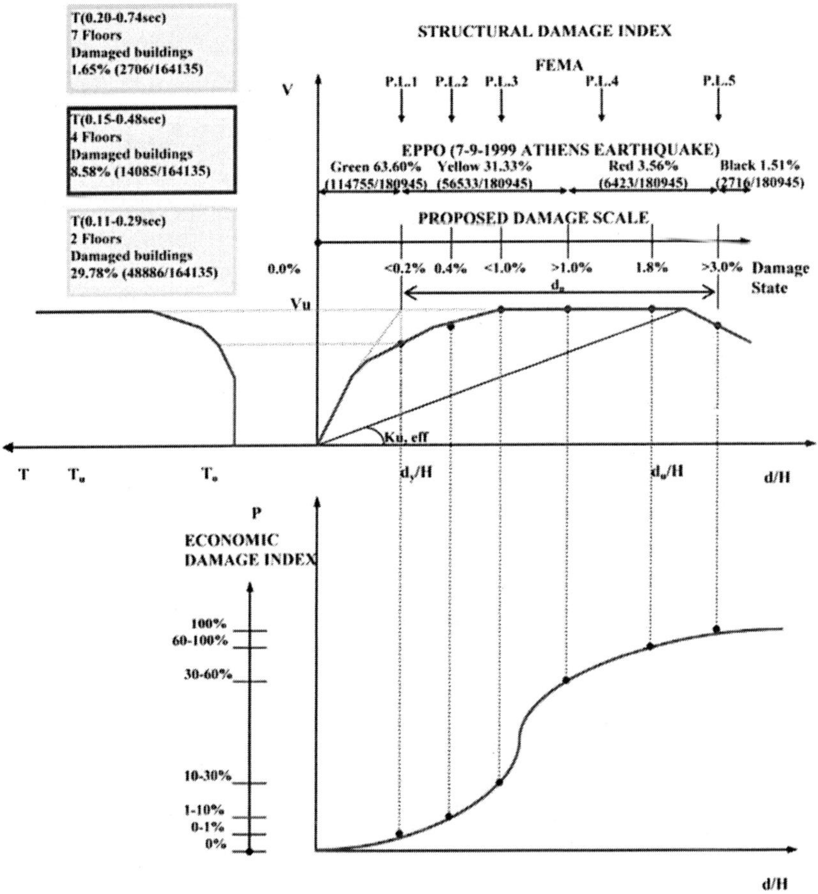

Figure 13. Correlation of the performance levels of the proposed damage scale with the fundamental period and a "primary vulnerability curve".

6. SUMMARY AND CONCLUSIONS

In the current research the fundamental period of vibration is calculated for several reinforced concrete building types according to existing simple relationships. The typical building typologies are representative of the materials, the seismic codes and the construction techniques of Southern Europe. They also represent groups of existing damaged buildings in several degree, type and extent of a dataset derived from post-earthquake surveys which took place in an extended region of Attica in Greece. The damage database with thousands of recorded buildings created after the occurrence of the Athens (7-9-1999) near field seismic event in an extended urban region [5,6]. The estimated fundamental period of typical reinforced concrete (RC) structures is correlated with the recorded seismic damage. The advantage of the research is that the empirical vulnerability assessment is the most realistic as it is based on a real experiment in a physical scale 1:1. Important conclusions are drawn on the parameters (height, structural type, etc.) that influence the seismic response and the development of damage based on the wide database. In addition, a damage scale for the measurable recording, beyond the qualitative characterization of seismic damage in Greek post-earthquake surveys, is presented wherein the performance levels are defined according to the physical description of the seismic damage and, as well, in terms of structural and economic damage index. A correlation is fulfilled between the estimated fundamental periods of RC buildings with the different damage states of the proposed scale, FEMA performance levels and the analysis results of the database.

In addition, a correlation analysis is fulfilled between the elastic acceleration spectra of artificial seismic ground motions, accelerograms and the evaluated fundamental periods for the considered in Greece and generally in Southern Europe as "low", "medium" and "high" height buildings along with the elastic spectrum of the first (1959) and the contemporary (2003) Greek Seismic Code. It is concluded that in the most records the peak ground accelerations of the 7th of September Athens earthquake were between the lower and the upper bound of the estimated fundamental periods based on experimental formulas [20] for all ("low"-"medium"-"high") RC building types with regular infills (n-normal) and with ground levels without infill panels (p-pilotis). A disparity is noticed for the estimated fundamental period according to EC8 provisions of "high" height buildings regarding the referring earthquake. Noticeable is the difference between seismic demand of the seismic elastic spectrum of the first (1959), the contemporary (2003) Greek Seismic Code and the values of peak ground accelerations of the several Athens earthquake records.

From the analysis results it occurs that the majority of buildings that developed several degree, type and extent of damage belong to "low" height buildings with estimated values close to the peak ground accelerations of the Athens earthquake ground motions. The last conclusion is connected to the fact that these buildings belong to the meizoseismal area of the northern suburbs close to the epicentral area (e.g. Ano Liosia, Aharnes, Fyli, Menidi, Ilion, Filadelpheia, Kamatero, Zefyri) where the most serious damages were observed. In several of these regions the macroseismic intensity reached IX.

The aspect that the causes of the developed damages are owed to a single parameter would be very simplified. In general, damage displayed significant differentiation from place to place, as well as a peculiar geographic distribution. Based on geological, tectonic and morphological characteristics of the affected area and on the elaboration of damage recordings for intensity evaluation, it has been suggested that intensity distribution was the result of the combination of a number of parameters. On the one hand, the parameters are the strike of the seismogenic fault, seismic wave directivity effects and an old NNE ± SSW tectonic structure, and they are also responsible for the maximum intensity arrangement in two perpendicular directions ESE ± WNW and NNE ± SSW. On the other hand, site foundation formations, old tectonic structures buried under recent formations and morphology are the parameters that differentiated intensities within the affected area [43]. Generally, the unlike damage distribution of the 1999 Athens earthquake reflected the destructive combination of two factors: the source directivity and the site effect [44]. Surprisingly heavy damage occurred on the eastern bank of the Kifissos River canyon. A number of these buildings suffered partial or total collapse, while many others were severely damaged. Despite the particular geometry of the slope of about 60 m in height that caused significant motion amplification, topography effects alone cannot explain the disparity in damage distribution which is characterized by a rather uniform structural quality. Soil stratigraphy and material heterogeneity on the topographic aggravation of surface ground motion played important role. Several simulations showed that topographic effects are substantial only within about 50 m from the canyon ridge, materializing primarily because of the presence of relatively soft soil layers near the surface of the profile. The results showed that both topography and local soil conditions significantly affected the spatial variability of seismic motion [45,46].

REFERENCES

1. R. K. Goel and A. K. Chopra, "Period Formulas for Moment—Resisting Frame Buildings," Journal of Structural Engineering, Vol. 123, No. 11, 1997, pp. 1454-1461.

2. R. K. Goel and A. K. Chopra, "Period Formulas for Concrete Shear Wall Buildings," Journal of Structural Engineering, Vol. 124, No. 4, 1998, pp. 426-433.

3. H. Crowley, "Periods of Vibration for DisplacementBased Assessment of RC Buildings," MSc Dissertation, ROSE School, Pavia, 2003.
4. H. Crowley and R. Pinho, "Period—Height Relationship for Existing European Reinforced Concrete Buildings," Journal of Earthquake Engineering, Vol. 8, No. 1, 2004, pp. 93-119.
5. A. K. Eleftheriadou, "Contribution to the Seismic Vulnerability Assessment of Reinforced Concrete Structures (in Greek)," Ph.D. Thesis, Department of Civil Engineering, Democritus University of Thrace, Xanthi, 2009.
6. A. K. Eleftheriadou and A. I. Karabinis, "Seismic Vulnerability Assessment of Buildings Based on the Near Field Athens (7-9-1999) Earthquake Damage Data," International Journal of Earthquakes and Structures, Vol. 3, No. 2, 2012, pp. 117-140.
7. A. K. Eleftheriadou and A. I. Karabinis, "Evaluation of Damage Probability Matrices from Observational Seismic Damage Data," International Journal of Earthquakes and Structures, Vol. 4, No. 3, 2013, in press.
8. H. Crowley and R. Pinho, "Revisiting Eurocode 8 Formulae for Periods of Vibration and Their Employment in Linear Seismic Analysis," Earthquake Engineering & Structural Dynamics, Vol. 39, No. 2, 2010, pp. 223-235.
9. M. J. N. Priestley, "Displacement-Based Seismic Assessment of Reinforced Concrete Buildings," Journal of Earthquake Engineering, Vol. 1 No. 1, 1997, pp. 157-192.
10. S. Glaister and R. Pinho, "Development of a Simplified Deformation-Based Method for Seismic Vulnerability Assessment," Journal of Earthquake Engineering, Vol. 7, No. 1, 2003, pp. 107-140.
11. G. M. Verderame, I. Iervolino and G. Manfredi, "Elastic Period of Sub-Standard Reinforced Concrete Moment Resisting Frame Buildings," Bulletin of Earthquake Engineering, Vol. 8, No. 4, 2010, pp. 955-972.
12. Applied Technology Council (ATC), "Tentative Provisions for the Development of Seismic Regulations for Buildings," Report No. ATC3-06, Palo Alto, 1978.
13. Structural Engineers Association of California, "Recommended Lateral Force Requirements," SEAOC, San Francisco, 1988.

14. D. Gilles and G. McClure, "Development of a Period Database for Buildings in Montreal Using Ambient Vibrations," Proceedings of the 14th World Conference on Earthquake Engineering, Beijing, 12-17 October 2008, Paper No. 12-03-0016, 8 p.

15. Applied Technology Council (ATC), "Minimum Design Loads for Buildings and Other Structures," ASCE/SEI 7-05, Structural Engineering Institute of the American Society of Civil Engineers, Reston, 2006.

16. CEN, "Eurocode 8: Design of Structures for Earthquake Resistance. Part 1: General Rules, Seismic Actions and Rules for Buildings," European Standard EN 1998-1:2004, Comité Européen de Normalisation, Brussels, 2004.

17. P. Ricci, G. M. Verderame and G. Manfredi, "Analytical Investigation of Elastic Period of Infilled RC MRF Buildings," Engineering Structures, Vol. 33, No. 2, 2011, pp. 308-319.

18. Uniform Building Code (UBC-97), "Structural Engineering Design Provisions," International Conference of Building Officials, Whittier, 1997, p. 492.

19. SEAOC, "Recommended Lateral Force Requirements," 6th Edition, Seismology Committee, Structural Engineers Association of California, Sacramento, 1996.

20. A. I. Karabinis, "Contribution to the Research of Dynamic Characteristics for Reinforced Concrete Structures," Ph.D. Thesis, Department of Civil Engineering, Democritus University of Thrace, Xanthi, 1986.

21. Institute of Engineering Seismology and Earthquake Engineering (ITSAK)—Aristotle University of Thessaloniki (AUTh), "Athens Earthquake: Assessment of Vulnerability in the Disaster Area and Correlation to the Real Distribution of Buildings Damage after the Earthquake," Research Project, Earthquake Planning and Protection Organization (EPPO), Thessaloniki, 2004.

22. A. K. Eleftheriadou and A. I. Karabinis, "Development of Damage Probability Matrices Based on Greek Earthquake Damage Data," Journal of Earthquake Engineering & Engineering Vibration, Vol. 10, No. 1, 2011, pp. 129-141.

23. Z. Schenková, V. Schenk, I. Kalogeras, R. Pichl, P. Kottnauer, C. Papatsimba and G. Panopoulou, "Isoseismal Maps Drawing by the

Kriging Method," Journal of Seismology, Vol. 11, No. 1, 2007, pp. 121-129.

24. L. Hutchings, E. Ioannidou, W. Foxall, N. Voulgaris, J. Savy, I. Kalogeras, L. Scognamiglio and G. Stavrakakis, "A Physically Based Strong Ground-Motion Prediction Methodology; Application to PSHA and the 1999 Mw = 6.0 Athens Earthquake," Geophysical Journal International, Vol. 168, No. 2, 2007, pp. 659-680.

25. National Technical Chamber of Greece (NTCG), "PreEarthquake Reinforcement of Existing Buildings (in Greek)," National Programme for Earthquake Management of Existing Buildings, Earthquake Planning & Protection Organization, Athens, 2006.

26. A. K. Eleftheriadou and A. I. Karabinis, "Damage Probability Matrices Derived from Earthquake Statistical Data," Proceedings of the 14th World Conference on Earthquake Engineering, Paper No. 07-0201, Beijing, 12-17 October 2008.

27. A. K. Eleftheriadou and A. I. Karabinis, "Seismic Vulnerability Assessment with Damage Probability Matrices," Proceedings of the 3rd Greek Conference on Earthquake Engineering and Technical Seismology, Paper No. 2108, Athens, 2008.

28. A. K. Eleftheriadou and A. I. Karabinis, "Seismic Damage Scales in Reinforced Concrete Structures," Technika Chronika, Scientific Journal of the Technical Chamber of Greece, Vol. 3, No. 1, 2010, pp. 41-60.

29. T. Rossetto and A. Elnashai, "Derivation of Vulnerability Functions for European-Type RC Structures Based on Observational Data," Journal of Engineering Structures, Vol. 25, No. 10, 2003, pp. 1241-1263.

30. M. Rota, A. Penna and C. L. Strobbia, "Processing Italian Damage Data to Derive Typological Fragility Curves," Soil Dynamics and Earthquake Engineering, Vol. 28, No. 10-11, 2008, pp. 933-947.

31. P. Sarabandi, D. Pachakis, S. King and A. Kiremidjian, "Development of Empirical Building Performance Functions Data from Past Earthquakes," Proceedings of ICASP- 9, San Francisco, 6-9 July 2003, pp. 629-635.

32. Applied Technology Council (ATC), "Earthquake Damage Evaluation Data for California," ATC-13 Report, Redwood City, 1985.

33. A. J. Kappos, K. Pitilakis, K. Morfidis and N. Hatzinikolaou, "Vulnerability and Risk Study of Volos (Greece) Metropolitan Area," Proceedings of the 12th ECEE, London, Paper No.74, 9-13 September 2002.

34. A. J. Kappos, "Seismic Vulnerability and Risk Assessment of Urban Habitat in Southern European Cities," Proceedings of the Urban Habitat Constructions under Catastrophic Events Workshop (COST C26), Prague, 30-31 March 2007, pp. 115-129.

35. S. Tesfamariam and M. Saatcioglu, "Risk-Based Seismic Evaluation of Reinforced Concrete Buildings," Earthquake Spectra, Vol. 24, No. 3, 2008, pp. 795-821.

36. A. I. Karabinis and A. K. Eleftheriadou, "Vulnerability Assessment Derived from Earthquake Damage Data," Proceedings of the ECCOMAS Thematic Conference on Computational Methods in Structural Dynamics and Earthquake Engineering, Rethymno, Paper No.1264, 13- 16 June 2007.

37. Earthquake Planning and Protection Organization (EPPO), "Guidelines and Forms for Immediate Post-Earthquake Screening of Reinforced Concrete Buildings (in Greek)," Athens, 1997.

38. Ministry of Public Works, Earthquake Planning and Protection Organization, "Post-Earthquake Inspection of Reinforced Concrete Buildings (in Greek)," Athens, 1984.

39. National Technical Chamber of Greece (NTCG), "Vulner ability Assessment of Buildings (in Greek)," Final Report, Technical Team No.I.2, National Programme for Earthquake Management of Existing Buildings, Earthquake Planning & Protection Organization, Athens, 2001.

40. A. J. Kappos, K. C. Stylianidis and K. Pitilakis, "Development of Seismic Risk Scenarios Based on a Hybrid Method of Vulnerability Assessment," Journal of Natural Hazards, Vol. 17, No. 2, 1998, pp. 177-192.

41. A. Ghobarah, "On Drift Limits Associated with Different Damage Levels," Proceedings of International Workshop on Performance-Based Seismic Design, Department of Civil Engineering, McMaster University, Bled, 28 June-1 July 2004.

42. R. Foltz, "Estimating Seismic Damage and Repair Costs," MAE Center Project CM-4, The Citadel, Texas A&M, 2004, in press.

43. E. Lekkas, "The Athens Earthquake (7 September 1999): Intensity Distribution and Controlling Factors," Engineering Geology, Vol. 59, No. 3-4, 2001, pp. 297-311.

44. R. Roumelioti, A. Kiratzi and N. Theodulidis, "Stochastic Strong Ground—Motion Simulation of the 7 September 1999 Athens (Greece) Earthquake," Bulletin of the Seismological Society of America, Vol. 94, No. 3, 2004, pp. 1036-1052.

45. I. S. Kalogeras and G. N. Stavrakakis, "The Athens, Greece September 7th, 1999 Earthquake: Strong Motion Data Processing (7/9/1999-31/3/2000)," National Observatory of Athens, Geodynamic Institute, Athens, 2001.

46. G. Gazetas and Collaborators, "Computational and Experimental Assessment of Strong Motion within the Meizoseismal Area of Parnitha, 7-9-99, Earthquake (in Greek)," Technical Report, Earthquake Planning and Protection Organization, Athens, 2001, pp. 1-207.

8

CHAPTER

PUSHOVER ANALYSIS OF LONG SPAN BRIDGE BENTS

Vitaly Yurtaev[1,2] and Reza Shafiei[1,2]

[1] *Moscow State University of Railways, Department Of Civil Engineering, (MIIT), Moscow, Russia*

[2] *Washington State University, Pullman, WA, USA*

1. INTRODUCTION

It has been observed that most of the bridges damaged in earthquakes were constructed before 1971 and had little or no design consideration to seismic resistance. Since the 1971 San Fernando earthquake in California, the standards for earthquake design have been strengthened considerably, and bridge structural behavior has been more accurately evaluated. Since then, structural ductility, a crucial element for the survival of bridges under severe earthquakes has become a key consideration in structural analysis and design.

However, bridges that were constructed prior to 1971 are still in use and play important roles in our transportation systems, which may be susceptible to failure due to their structural deficiencies. To ensure safety and performance of these bridges, a seismic retrofit and strengthening program has been one of the major efforts of the Washington Department of Transportation and the Federal Highway Administration, aiming at improving seismic performance of older bridges. Retrofitting methods such as restrainers and column jacketing have proven to be effective in recent earthquakes. Techniques to retrofit other bridge members have also been developed such as soil anchors, footing retrofit involving increased plan dimension and reinforced overlay, construction of link beams, and system isolation and damping device.

The goal of seismic retrofit is to minimize the likelihood of structural failure while meeting certain performance requirements. This allows engineers to design repair strategies based on performance needs. As a consequence, some level of damage may be acceptable during a design-level earthquake. The California Department of Transportation (Caltrans) has required that bridge retrofits provide survival limit-state protection at seismic intensities appropriate for new bridges. This makes possible the proposition of efficient and effective strengthening measures with optimized retrofitting schemes, and the adoption of the plan that is the most economical for the acceptable damage level. One of the ways of implementation the retrofit program for the structures is providing a nonlinear static analysis.

Nonlinear static analysis under monotonically increasing lateral loading is becoming an increasingly popular tool for seismic performance evaluation of existing and new structures. Pushover analysis can be viewed as a method for predicting seismic force and deformation demands, which accounts in an approximate manner for the redistribution of internal forces occurring within the inelastic range of structural behavior. It is expected to provide information on many response characteristics that cannot be obtained from an elastic static or dynamic analysis. Pushover analysis is based on the assumption that the response of the structure can be related to the response of an equivalent SDOF system. This implies that the response is controlled by a single mode, and that the shape of this mode remains constant through the time history response. These assumptions are likely to be reasonable if the structure response is not severely affected by higher mode effects, or the structure has only a single plastic mechanism that can be detected by an invariant load pattern. The use of at least two load patterns that are expected to bound inertia force distribution is recommended. For structures that vibrate primarily in the fundamental mode, pushover analysis will very likely provide good estimates of global as well as local inelastic deformation demands. It will also expose design weaknesses that may remain hidden in an elastic analysis. Such weaknesses include storey mechanisms, excessive deformation demands, strength irregularity, and overloads on potentially brittle elements, such as columns and connections. On the negative side, the most critical is the concern that the pushover analysis may detect only the first local mechanism that will form in an earthquake and may not expose other weaknesses that will be generated when the structure's dynamic characteristics change after the formation of the first local mechanism.

2. OBJECTIVES OF THE ANALYSIS

The purpose of this research is to evaluate the displacement capacity of bents from a long span bridge. A three-dimensional nonlinear finite-element model of the bridge bents were developed to determine the inelastic response by performing nonlinear pushover analysis. Modal pushover analyses were carried out in the transverse direction. Detailed data of performance was collected and interpreted to use as a baseline in a parametric study. Separate parametric study was carried out on a single column within this bridge in order to locate appropriate plastic hinge locations. These results were then transferred to individual bents, where multiple columns were modeled based on the results from the single column parametric study.

3. DESCRIPTION OF THE BRIDGE

The bridge is located at the Primary State Highway No.1, Seattle Freeway Ravenna Boulevard Overcrossing North Bound. Figure 1 shows an aerial view of the bridge. The North Bound Bridge is the first bridge from below shown in the picture. The bridge length is 1310 ft back to back of pavement seats and consists of twenty spans. Plan and elevation views are shown in Figure 2 and Figure 3.

The superstructure is composed of pre-tensioned concrete beams. Each span includes twelve girders, and the general girder cross-section varies for each span. This complicates the calculation of the total mass of the superstructure. In order to simplify the procedure, individual span cross-sections were drawn in AutoCAD. The sections can be found is Figure 4 relating them to the spans they are assembled for. A table with calculated weight and length for each span can be found in the Appendix. Overlaid on top of the girders is a 5 in thick, approximately 60 ft wide reinforced concrete deck slab.

Figure 1. Aerial view of the North Bound bridge

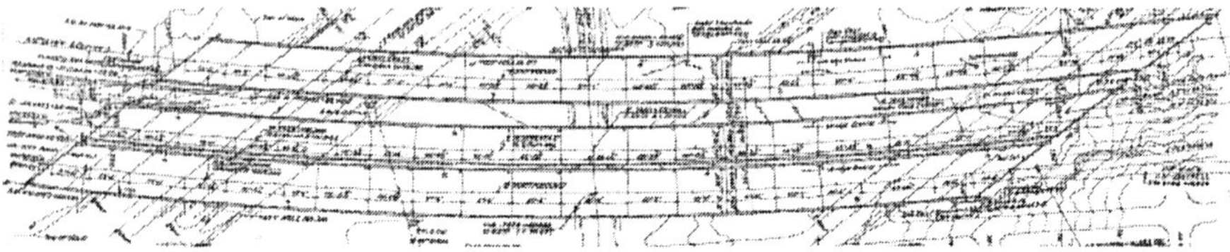

Figure 2. Plan view of the North Bound Bridge

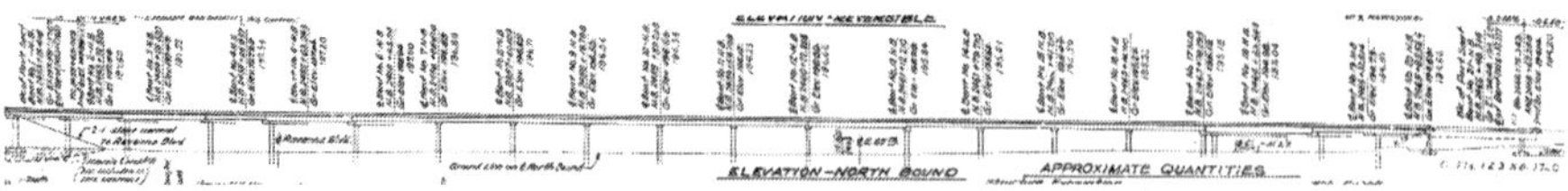

Figure 3. Elevation view of the North Bound Bridge

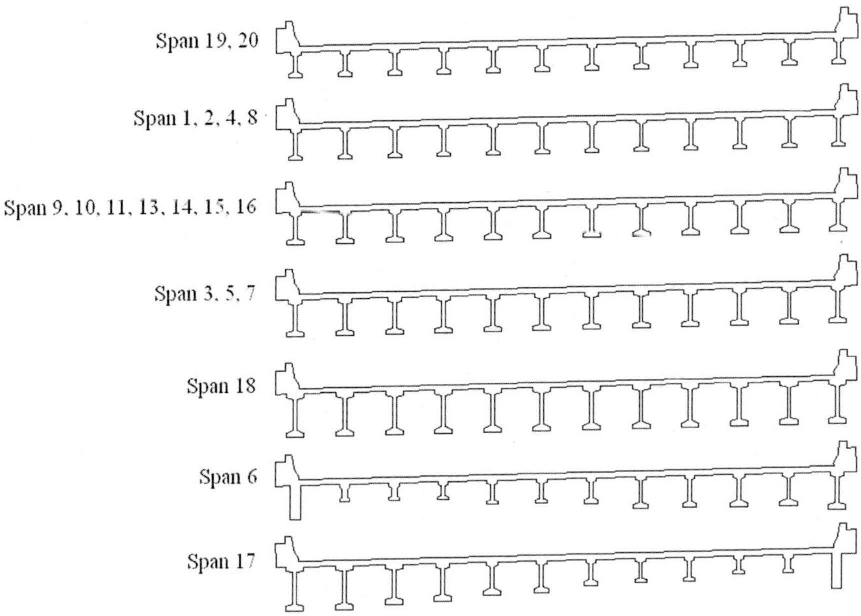

Figure 4. Superstructure Sections

There are a total of 19 bents in the bridge. Five are 6-column bents (#1-6), three are 7-column bents (#18-20) and ten are 4-column bents (#7-17). The cross-beam plans for the three types of bents are shown in Figure 5. Each bent has a unique elevation above the ground. Also, because of the curved shape of the bridge, each bent has a slight rotation in the vertical direction. Consequently, there is column height variation within each bent. The various column height values can be found in the Appendix.

At each bent, a 3x4.6 ft crossbeam transversely connects the columns. Figure 6 below shows the geometry and steel reinforcement. The length of the beams varies for each bent, which can be found in the Appendix. The steel reinforcement consists of nine No. 10 bars located at the top and at the bottom of each crossbeam. Two No. 5 bars are located at the side edges and run longitudinally along the crossbeam. For shear reinforcement, No. 5 stirrups are spaced evenly along each member.

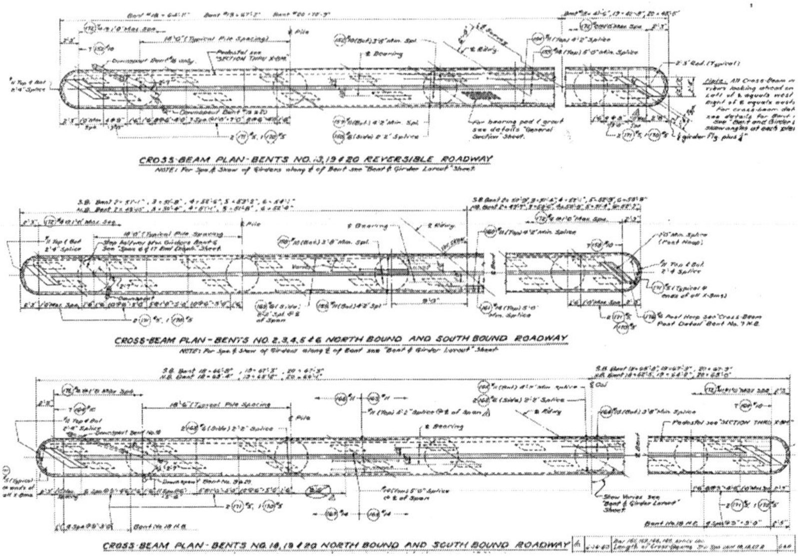

Figure 5. Cross-Beam Plan for Bents

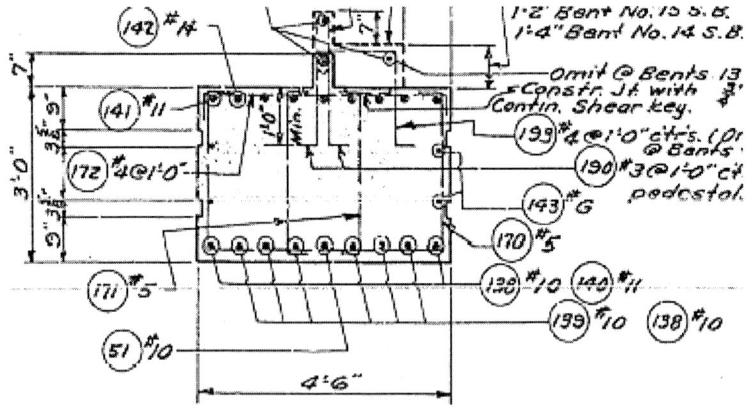

Figure 6. Section Thru Cross-Beam

The columns are spaced at 18 ft centerline to centerline. Each column is hollow with an outer diameter of 48 in and a wall thickness of 5 in. Twelve evenly spaced No. 5 bars provide the longitudinal reinforcement within each column. The columns also include twelve No. 3 steel cables each post-tensioned initially to 61 kips. Transverse reinforcement is provided by No. 2 spiral hooping spaced at 6 in on center. Figure 7 shows the plan column section. The columns are extended approximately 27 ft into the ground to act as piles.

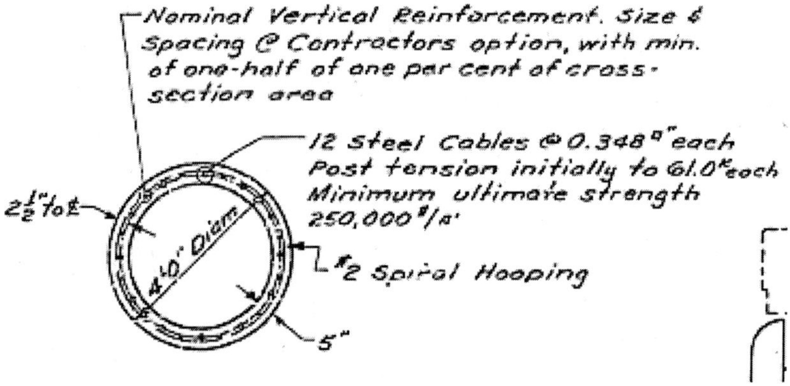

Figure 7. Plan Column Section

The columns and crossbeam were cast monolithically adding considerable rigidity to each bent. Figure 8 shows the elevation view of a typical bent. Further, the top 4 ft of each column is filled with class A concrete. This fill is further reinforced with sixteen No. 8 bars longitudinally, and No. 3 hoops spaced at 12 in transversely. In this section of the column, the hollow column is transversely reinforced with No. 2 spiral hooping spaced at 3 in over center. Figure 9 shows a typical pile.

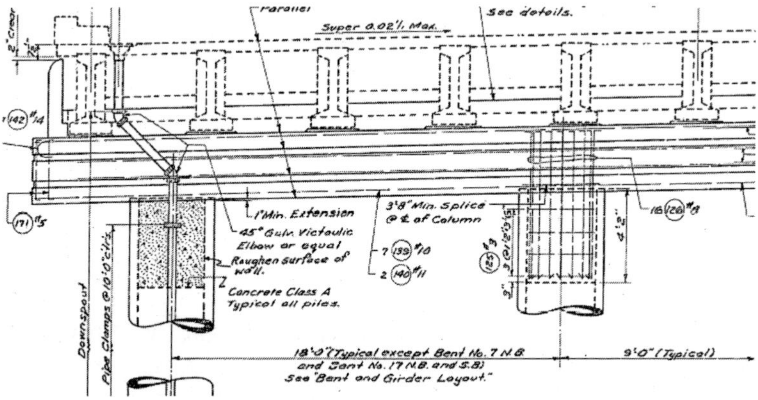

Figure 8. Elevation View of Bents

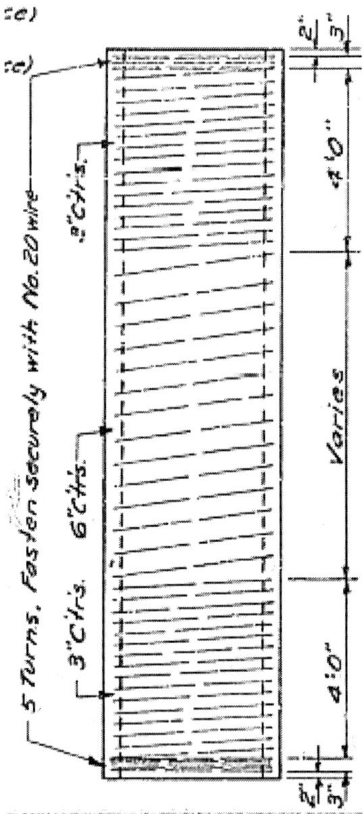

Figure 9. Typical Pile

4. MODELING OF THE BENTS

A spine model of each bent is created in the finite element program SAP2000. Line elements can behave three-dimensionally in the form of beam, beam-column elements and springs. The superstructure is represented as a distributed dead load which represents the dead weight of the superstructure based on tributary length of related spans for each bent. A table in the Appendix provides the distributed load values used in the analysis for each bent. The soil-structure interaction is represented by springs. In order to capture nonlinear behavior of the columns, plastic hinges were defined at maximum moment points. The general model is represented in 3D in Figure 10.

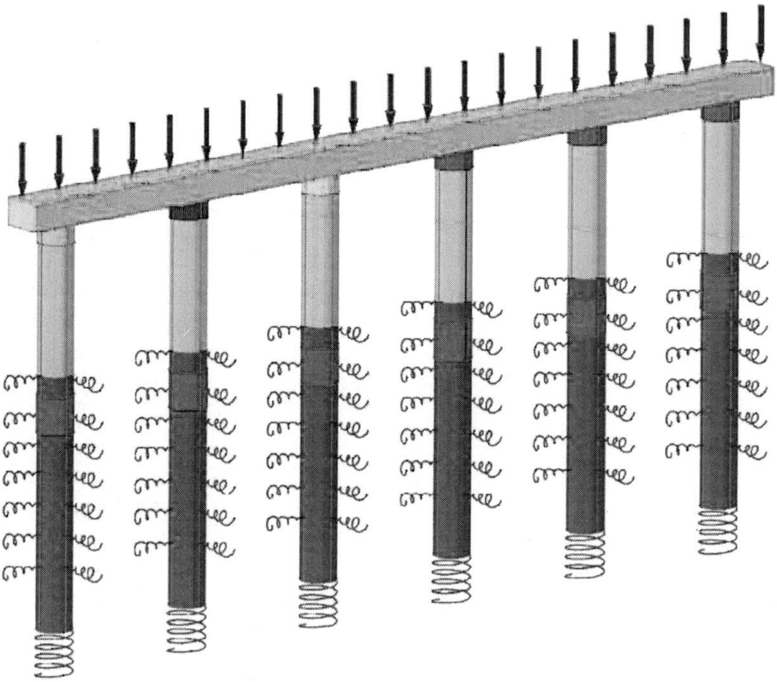

Figure 10. Model of Six-Column Bents

The cross-sections of the cap beam and the column were accurately modeled by using the subprogram offered in SAP2000 called Section Designer. Section Designer lets the user draw the shape of the cross-section and also include the steel reinforcement. Figure 11 shows the drawn sections used in the analysis.

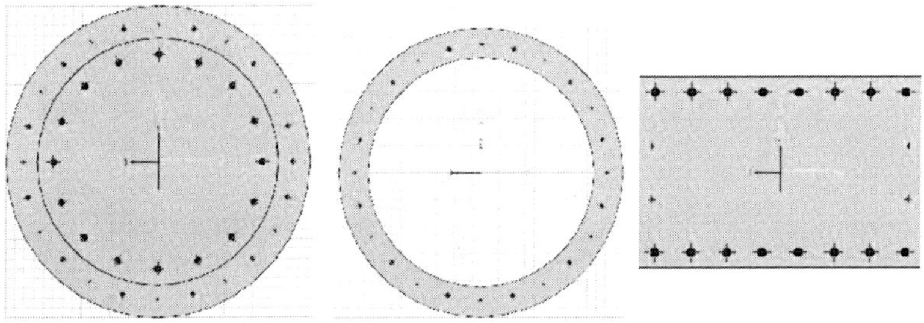

Figure 11. Bent Element Cross-Sections

Piles

The columns were considered fixed in the cap beam. Nonlinear springs along the pile shafts were used to model the resistance provided by the surrounding soil. The L-Pile software (2002) was used to compute the P-Y curves, based on the stiff sand soil model with free water at 15 depths.

To build an exact computer model of a structure beard against underground elements-piles it is necessary to know how interaction between soil and a pile can be simulated, to get more precise result of the analysis. The p-y curves is a strait interpretation of the relation between deflection of an element and soil pressure on a particular depth. The pressure from the soil on the element is distributed within certain length which depends on the number of springs assigned to it Figure 12.

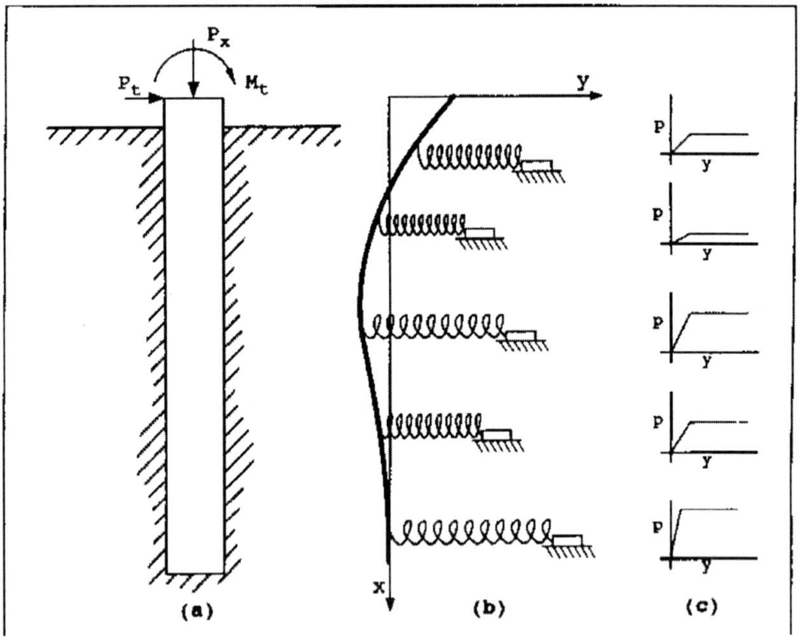

Figure 12. Model of laterally loaded pile

A physical definition of the soil resistance *p* is given in Figure 13. There was made an assumption that the pile has been installed without bending so the initial soil stresses at the depth x_i are uniformly distributed as shown in Figure 13*b*. If the pile is loaded laterally so that a pile deflection y_i occurs at the depth x_i the soil stresses will become unbalanced as shown in Figure

13*c*. Integration of the soil stresses yielding the soil resistance p_i with units F/L equation 1.

$$p_i = E_s y_i \tag{1}$$

where,

E_s – a parameter with the units F/L^2, relating pile deflection y and soil reaction p.

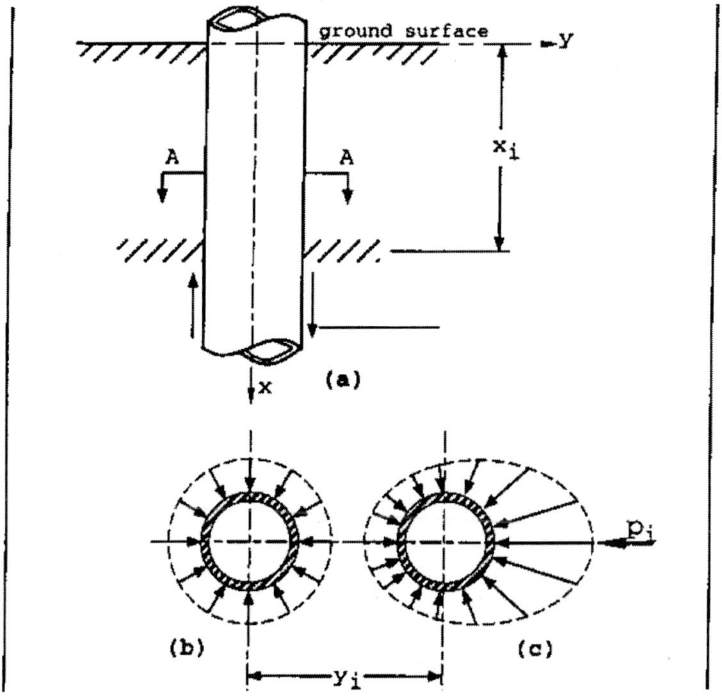

Figure 13. Definition of p and y as Related to Response of a Pile to Lateral Loading

Once the p-y curves at various depths of the pile have been obtained, a force-displacement relationship can be calculated by multiplying p with the tributary length of the pile between springs. Figure 14 shows a bilinearization of the force-displacement relationship at different depth based on the data retrieved from LPILE single pile analysis. These results were used to define multi-linear elastic links (springs) in SAP2000 in order to represent the SSI of the piles. The piles of all bents were assumed to extend 27 ft under the ground, so all bents had the same pile modeling.

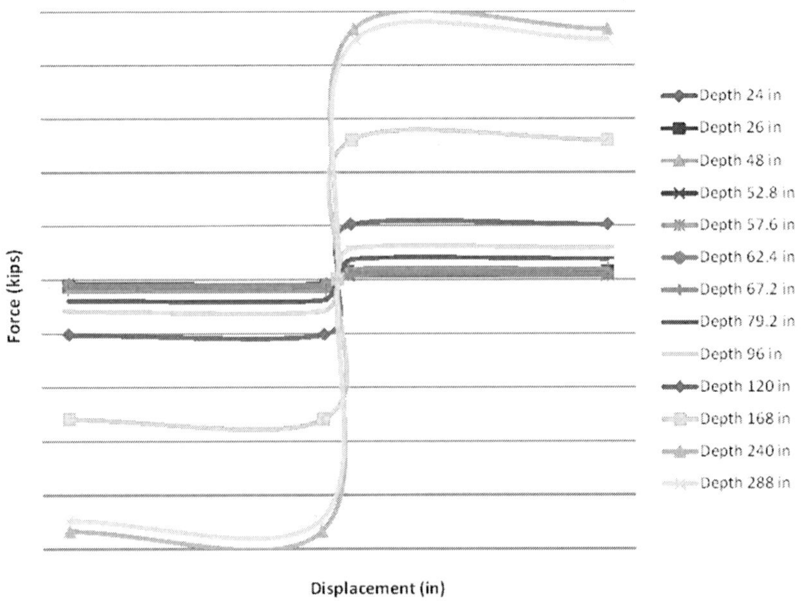

Figure 14. Bilinearized Force-Displacement of SSI at Different Depths

5. PLASTIC HINGE

It is well known that well-confined concrete structures can deform inelastically without significant strength loss through several cycles of response. Ductility describes such ability of structures, which is often defined as the ratio of deformation at a given response level to the deformation at yield response. Commonly used ductility ratios include displacement ductility, curvature ductility and rotation ductility. In the software of XTRACT, developed by Imbsen & Associates Company (2002) with the capability of analyzing structural cross sections, curvature ductility can be calculated for a given section and are defined in Equation 2 (Paulay and Priestley, 1992).

$$\mu_\varphi = \frac{\varphi_u}{\varphi_y} = \frac{\varphi_p + \varphi_y}{\varphi_y} \tag{2}$$

in which ϕ_y is yield curvature, ϕ_p is plastic curvature, and ϕ_u is summation of yield curvature and plastic curvature that presents the ultimate curvature capacity of a section.

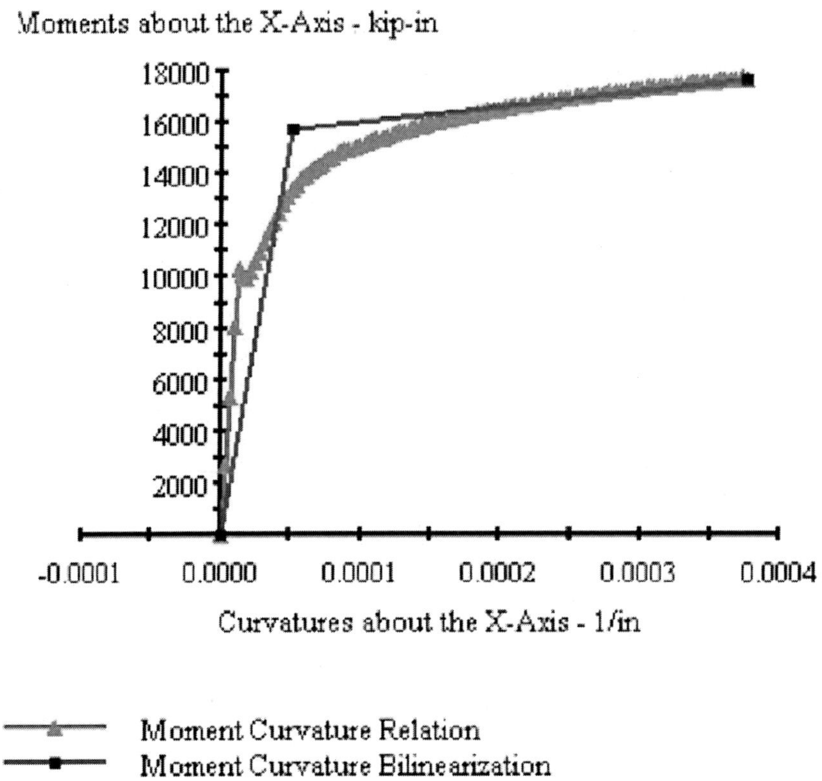

Figure 15. Bilinearization of the Moment–Curvature Curve for Hollow Column

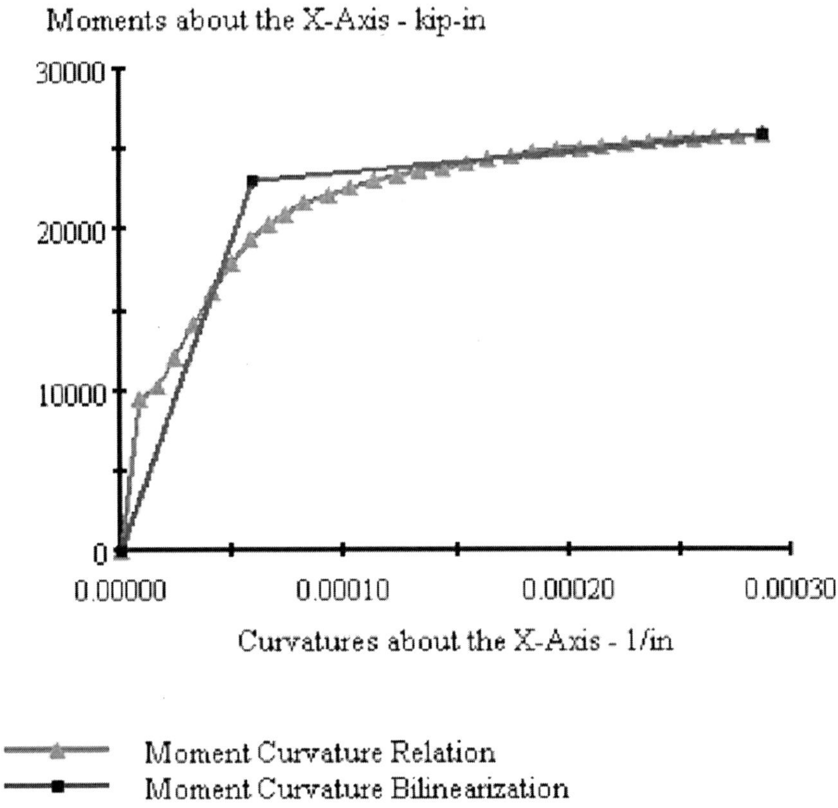

Figure 16. Bilinearization of the Moment–Curvature Curve for Filled Hollow Column

Figure 15 and Figure give a moment-curvature diagram for the column sections in the North Bound Bridge, calculated by the XTRACT. Curvature properties are section dependent and can be determined by numerical integration methods. Input data of a cross-section include nonlinear material properties of concrete and steel, and the detailed configuration of the section. For the North Bound bridge, all the columns have the identical section dimension, however, the moment-rotation relationships may not be the same because of the different axial loads.

Hinge length

The plastic hinge length for piles depends on whether the hinge is located at the pile/deck interface or is an in-ground hinge. For prestressed piles where the solid pile is embedded in the deck, the plastic hinge length at the pile/deck interface can be taken as (PIANC):

$$L_p = 0.5\, D_p \tag{3}$$

where,

D_p – diameter of a pile

For in-ground hinges, the plastic hinge length depends on the relative stiffness of the pile and the foundation material. Because of the reduced moment gradient in the vicinity of the in-ground hinge, the plastic hinge length is significantly longer there. In this report CALTRANS interpretation of in-ground hinges for a non-cased pile shaft was used. Figure 17 describes the calculation steps provided by CALTRANS.

$$L_p := D_e + 0.6 \cdot L$$

L_p = Plastic hinge length (in)

L = Length of pile shaft from ground surface to point of contraflexure above ground (in)

D_e = Diameter or least cross sectional dimension (in)

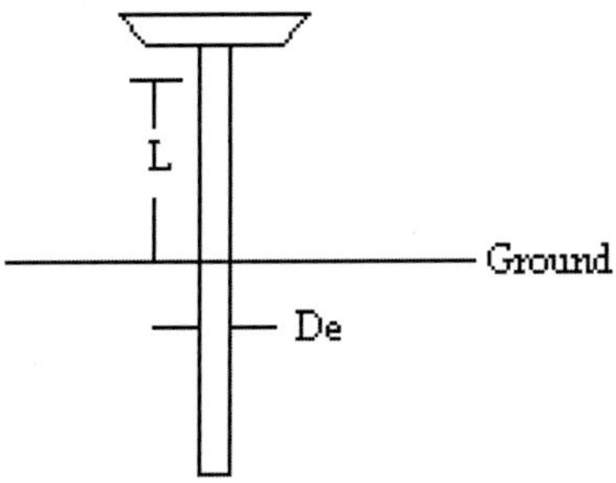

Figure 17. In-Ground Hinge Length

Hinge location

In order to locate the plastic hinge locations, a separate push over analysis was run on single column. Figure 18 shows the single column element modeled in SAP2000. Top of the column is restrained against rotation to represent the rigid connection between the column and the deck. The SSI is represented by links just as discussed for general bents. The pin connection at the bottom of the pile restricts the pile from vertical movement.

Figure 18. Single Column Finite Element Stick Model

Figure 19 provides the moment diagram of the above column/pile under horizontal loading. The diagram has two points of maximum moment. The plastic hinge should be placed at these locations in order to represent the most conservative nonlinear behavior of the column/pile.

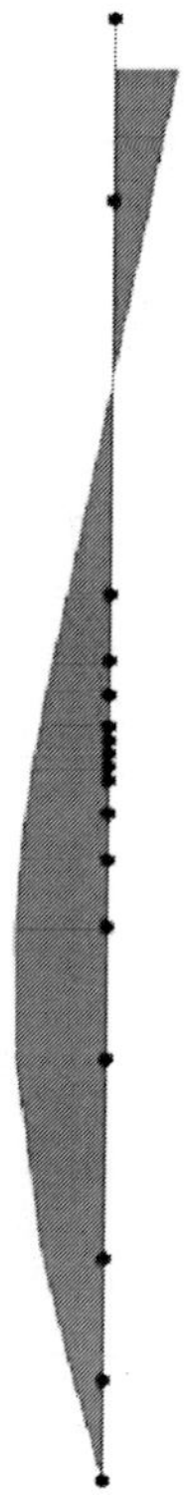

Figure 19. Moment Diagram of Single Column under Horizontal Load

The placement of the first hinge should be at the column/bent connection as expected before. The second hinge has to be place under the ground, but the location of maximum moment in that area changes in a pushover analysis. A parametric study was run in order to locate the worst location for an in-ground hinge. The placement of the in-ground hinge was varied for multiple pushover analysis. Figure 20 shows the results of this parametric study, where the hinge depth below ground level is compared to column top displacement capacity. The plot in Figure 20 shows that placing the hinge 20% of pile length under the ground would give a displacement capacity of 2.25 in, which is less than any other location. Figure 21 shows the placement of the plastic hinges in four column bent.

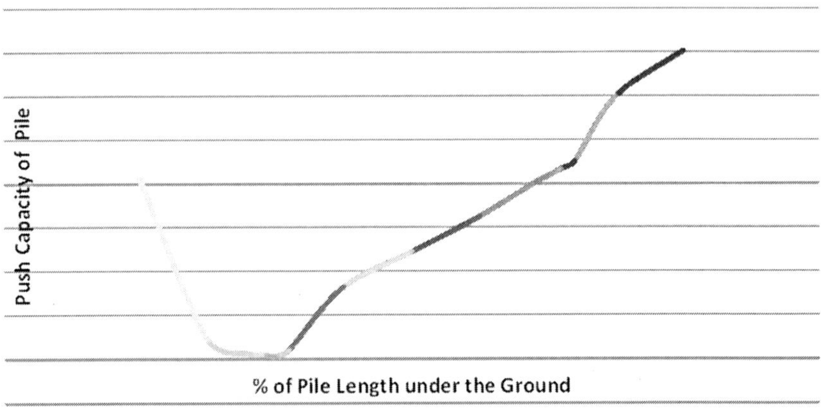

Figure 20. Single Column Parametric Study Results

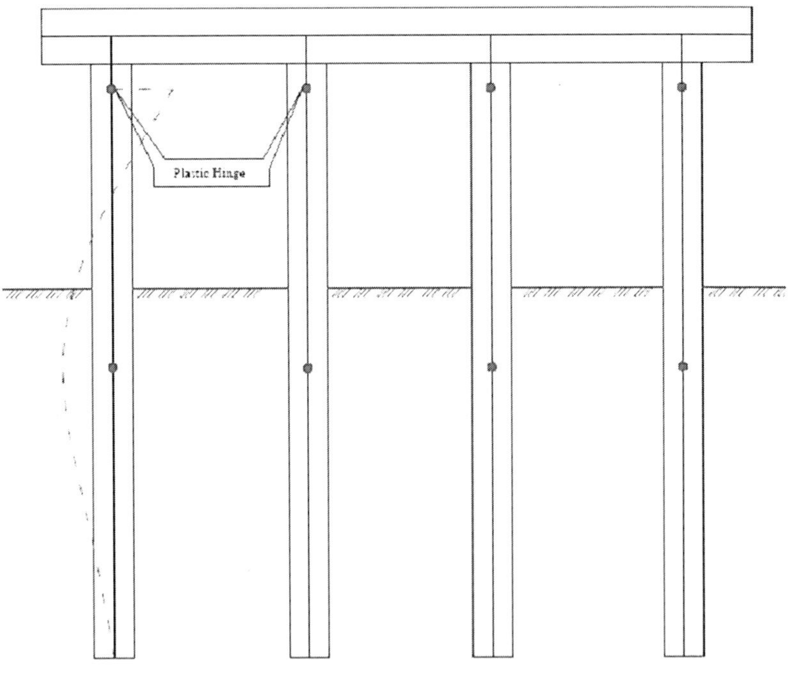

Figure 21. Location of the Plastic Hinge

Plastic hinge property

The Manual of SAP2000 recommends a distributed plastic hinge model assuming 0.1 of element length as the plastic hinge length, but information on how to define distributed plastic hinge properties is not provided. In this research, a concentrated plastic hinge model is used with the assumption that plastic rotation will occur and concentrate at mid-height of a plastic hinge. Input hinge properties consist of the section yield surface, plastic rotation capacity, and acceptance criteria.

A plastic rotation, θ_p, can be calculated by the plastic curvature given the equivalent plastic hinge length L_p as shown in Equation 4.

$$\theta_p = \varphi_p L_p = L_p(\varphi_u - \varphi_y) \tag{4}$$

The plastic rotation is an important indicator of the capacity of a section to sustain inelastic deformation and is used in SAP to define column plastic hinge properties. FEMA 356 provides a generalized force-deformation relation model shown in Figure 22 for the nonlinear static analysis procedure, which is the defaulted model in SAP for the Axial-Moment hinge.

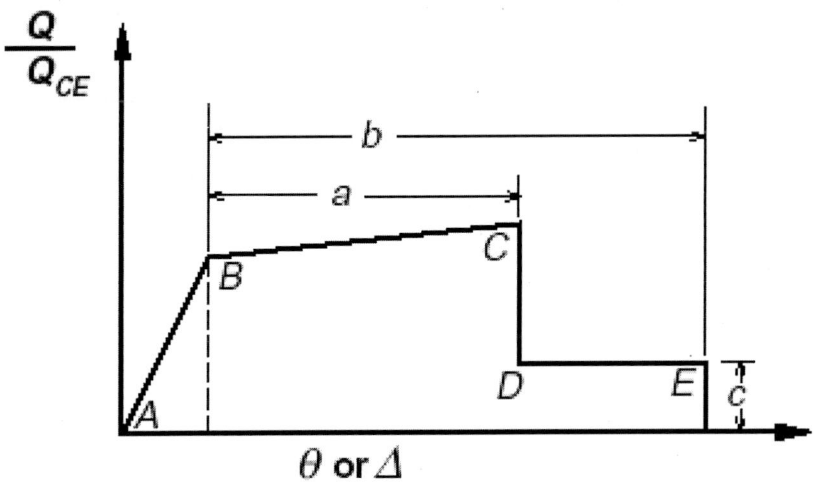

Figure 22. Generalized Force-Deformation Relations for Concrete Elements (FEMA-356)

Three parameters, *a*, *b* and *c* are defined numerically in FEMA-365, and are permitted to be determined directly by analytical procedures. The moment and rotation are normalized by yield moment and yield rotation respectively, i.e., $\frac{M}{M_y}$ and $\frac{\theta}{\theta_y}$. By default SAP will calculate the yield forces and the yield rotation based on reinforcement and section provided.

In Table 6-8 of FEMA 356, modeling parameters and numerical acceptance criteria are given for reinforced concrete columns in various categories. Columns investigated are all primary structural elements. A conforming transverse reinforcement is defined by hoops spaced in the flexural plastic hinge region less than or equal to $\frac{d}{3}$, and the strength provided by the hoops (*Vs*) being greater than three-fourths of the design shear. Thus, the category of the column is decided in Table 6-8 of FEMA 356, and values and relationship of the performance levels can be utilized.

In SAP, an absolute rotation value can overwrite the default value in defining a hinge property. The plastic rotation capacity angle, *a*, calculated with Equation 4-12 for a given column is at point C. The ultimate rotation angle, which is inputted as *b* in SAP, is taken as 1.5 times the plastic angle. It is indicated at point E, which defines a local failure at a plastic hinge. A larger value could be used to allow the structure to form a global failure due to instability.

The three discrete structural performance levels are Immediate Occupancy (IO), Life Safety (LS) and Collapse Prevention (CP) shown in Figure 23.

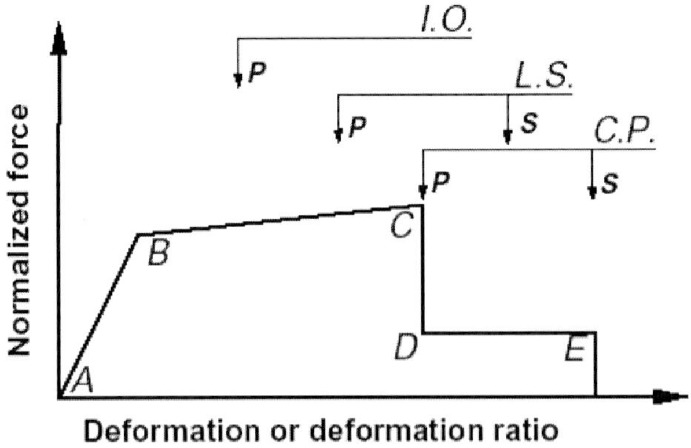

Figure 23. Performance Level on Generalized Force-Deformation Relations for Concrete Elements (FEMA-356)

The ultimate plastic hinge angle calculated by the XTRACT was taken as the Collapse Prevention level. Its value was indicated as "*a*" in Figure 1. The permissible deformation for the Life Safety performance level is taken as three quarters of the plastic rotation capacity "*a*".

The increase of moment strength at point C is taken as the over strength factor computed by XTRACT, ignoring the strength softening effect. The actual moment strength at point C is the product of the factor and the yielding moment. FEMA 356 defines a 0.2 residual strength ratio before plastic hinge eventually fails. Figure 24 presents moment-rotation curves for one of the columns.

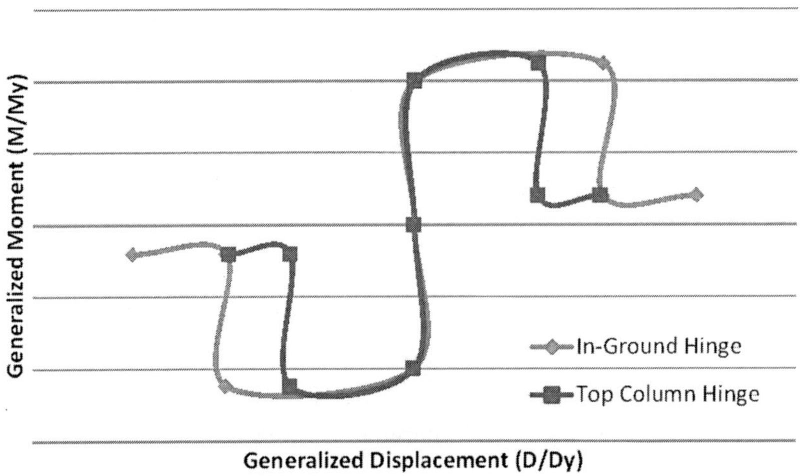

Figure 24. Moment-Rotation Relationship of the Columns

A concrete interaction surface was obtained from XTRACT for the frame hinges under combined bending and axial load. A generated interaction surface is shown in Figure 25.

6. RESULTS

The results of the pushover analysis are organized in the following pages. Each bent has two pages of results. The first page shows the general characteristics of the bent and the push over curve. The second page shows the results of the step by step push over analysis and the development of hinges. The first and the last bent on the plan are abutments, so they are not included in this report.

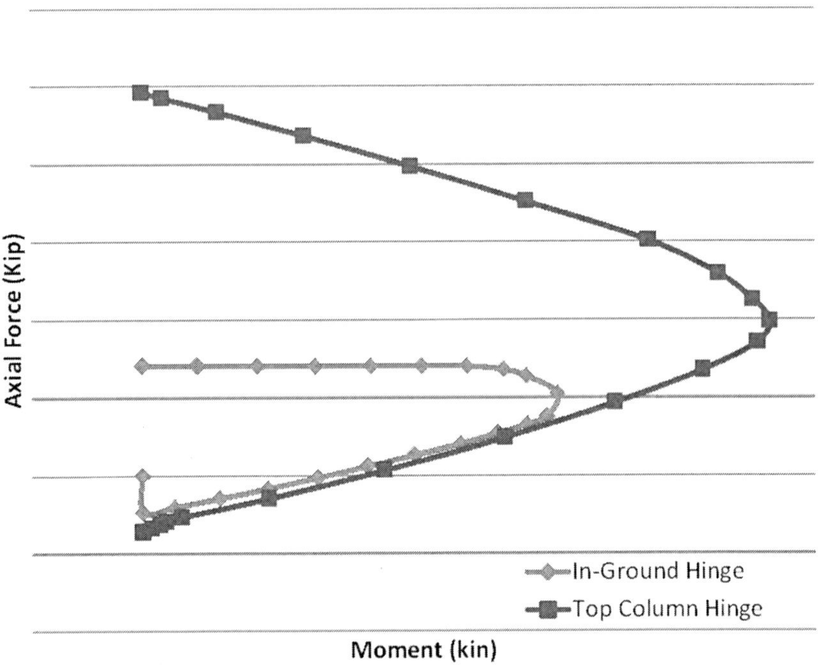

Figure 25. Axial Load-Moment Interaction Curve (Compression force is negative)

Maximum displacement capacity exhibited is 3.9 inches by the fifth bent. The lowest bent capacity is 2.08 inches by the second bent. The reason for this low displacement capacity in bent number two is because of its shorter columns compared to other bents.

Using two hinges per column creates a much lower capacity than using only one hinge at the top of the column. The shear force capacity stays almost the same, but the ductility reduces for all bents.

REFERENCES

1. FEMA 3562000table 68
2. M. J. N Priestley, Frieder Seible, Gian Michele. Seismic Design and Retrofit of Bridges, Wiley-IEEE, 19961996200250p.
3. B Pui-shum, Shing, National Science Foundation (U. S.), Tadaaki Tanabe, Nihon Gakujutsu Shinkōkai, National Science Foundation (U.S.), Nihon Konkurīto Kōgaku Kyōkai, Nihon Gakujutsu Shinkōkai, Nihon Konkurīto Kōgaku Kyōkai, Modeling of inelastic behavior of RC structures under seismic loads, ASCE Publications, 20012001425450p.
4. J. H Rainer, and A Vanselst, Dynamic properties of Lions Gate Bridge', Proc. Conf. on Dynamic Response of Structures: Instrumentation,.esting Methods and System Identi,cation, ASCE Specialty Conf., University of California, 1976243252
5. J. H Rainer, and G Pernica, Dynamic.esting of a Modern Concrete Bridge, National Research Council of Canada, 1979447
6. A Nishimura, and H Haya, Assessment of the structural integrity of bridge foundation by the impact vibration test', GEO-COAST, Yokohama, 1991719724
7. R Shepherd, and G. K Sidewell, Investigations of the dynamic properties of concrete bridges, 4th Australian Conf. on the Mechanics of Structures and Materials, University of Queensland, Brisbane, Australia, 1977261
8. B. M Dougla, C. D Drown, and M. L Gordon, Experimental dynamics of highway bridges', in G. Hart (ed.) Proc. Specialty Conf. on Dynamic Response of Structures, Experimentation, Observation, Prediction and Control, ASCE, New York, 1981698

9

CHAPTER

GEOTECHNICAL CHARACTERIZATION OF MINED CLAY FROM APPALACHIAN OHIO: CHALLENGES AND IMPLICATIONS FOR THE CLAY MINING INDUSTRY

Anthony R. Moran [1] and Hiroshan Hettiarachchi [2,*]

[1] CH2M Hill, 7927 Nemco Way, Suite 120, Brighton, MI 48116, USA

[2] Department of Civil Engineering, Lawrence Technological University, 21000 West Ten Mile Road, Southfield, MI 48075, USA

ABSTRACT

Clayey soil found in coal mines in Appalachian Ohio is often sold to landfills for constructing Recompacted Soil Liners (RSL) in landfills. Since clayey soils possess low hydraulic conductivity, the suitability of mined clay for RSL in Ohio is first assessed by determining its clay content. When soil samples are tested in a laboratory, the same engineering properties are typically expected for the soils originated from the same source, provided that the testing techniques applied are standard, but mined clay from Appalachian Ohio has shown drastic differences in particle size distribution depending on the sampling and/or laboratory processing methods. Sometimes more than a 10 percent decrease in the clay content is observed in the samples collected at the stockpiles, compared to those collected through reverse circulation drilling. This discrepancy poses a challenge to geotechnical engineers who work on the prequalification process of RSL material as it can result in misleading estimates of the hydraulic conductivity of the samples. This paper describes a laboratory investigation conducted on mined clay from Appalachian Ohio to determine how and why the standard sampling and/or processing methods can affect the grain-size distributions. The variation in the clay content was determined to be due to heavy concentrations of shale fragments in the clayey soils. It was also concluded that, in order to obtain reliable grain size distributions from the samples collected at a stockpile of mined clay, the material needs to be processed using a soil grinder. Otherwise, the samples should be collected through drilling.

Keywords: mined clay; shale; Appalachian Ohio; clay content; grain size distribution; sieve; analysis; hydrometer analysis; Atterberg limits; hydraulic conductivity; Recompated Soil Liners (RSL)

1. INTRODUCTION

Clayey soil found beneath a coal layer, which is also known as underclay, is an ideal construction material for Recompacted Soil Liners (RSLs) in landfills due to its low hydraulic conductivity and relatively low cost. Coal mines in Appalachian Ohio typically encounter large volumes of underclay due to the depositional nature of coal formations there. Clay mines make a profit and increase usable space at the mines by mining and selling the underclay to landfills that do not have sufficient clay borrow sources located within their property boundaries or permitted limits of waste. Landfills can save money utilizing the mined clay if the haul distance is short enough with respect to the expense of using additional geosynthetic materials.

The research described in this paper comes to light from experiences with underclay from Appalachian Ohio, which has shown some unique characteristics in its composition. The Ohio region, at the base of the Appalachian basin, is comprised of shale enriched clay deposits found beneath coal layers [1]. These clay deposits are classified with a layer number. The layering system is based on the cyclothemic depositions of the coal layers in this region. The deeper the clayey soil layer, the lower the number identification for the layer [1].

The solid waste division of the Ohio Environmental Protection Agency (OEPA) requires that RSL material be pre-qualified prior to use in construction [2]. This means that samples are analyzed for grain-size distribution, Atterberg limits, compaction characteristics and hydraulic conductivity prior to their acceptance for use. For hydraulic conductivity, remolded samples satisfy the OEPA requirement for the establishment of the RSL's hydraulic conductivity [2]. Clay mines typically pay for the cost of the drilling exploration and allow consultants onto their sites to monitor the drilling of potential RSL borrow sources. The method of collecting samples for laboratory testing through drilling as well as collecting samples from representative stockpiles are deemed acceptable.

Standard field sampling methods and laboratory sample preparation procedures, should not produce different grain-size distributions for the same material. An engineering mind would expect that a sample collected within the borrow location at a clay mine would classify the same and contain the same clay content as when the soils are excavated and hauled to a stockpile at a landfill and re-tested. But experience with the mined clay from Appalachian Ohio has revealed that this is not always the case, and the mined clay from Appalachian Ohio has shown drastic differences in particle size distribution depending on the sampling and/or laboratory processing methods. More particularly, the clay content of the material, *i.e.* particles finer than 0.002 mm as per AASHTO [3], has been shown to vary substantially. The research described in this paper was conducted to investigate how the sampling and/or laboratory processing methods could affect the grain-size distribution of mined clay from Appalachian Ohio.

2. MATERIALS AND METHODS

Underclay samples were collected from a location in Appalachian Ohio that was previously used for coal mining. This underlay layer was specifically identified as Number 5 clay [1]. The clay source was originally under approximately 60 feet of overburden.

2.1. Sample Collection

About 100 soil samples were collected at different depths in the clayey soil deposit using a hollow stem auger with reverse circulation drilling. Before transporting to the laboratory, soil samples were first visually classified in the field and then collected in one gallon ziplock bags and labeled in numerical order by collection. Samples retrieved by this method were observed to have flour-like consistency as can be seen in Figure 1(a). All soil samples collected by drilling visually appeared nearly identical. The drilling efforts spanned approximately three days at the proposed borrow location at the clay mine. The samples collected through reverse circulation drilling were designated as D-samples during this research.

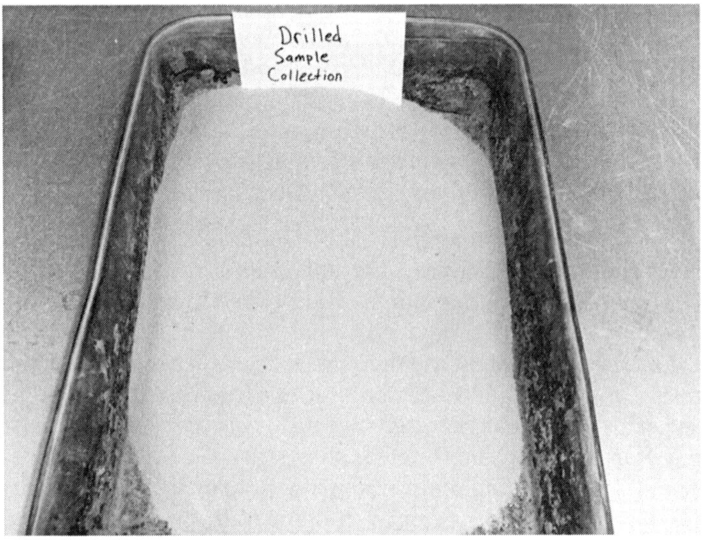

(a)

(b)

Figure 1. (**a**) Samples collected through drilling (D-samples); (**b**) Samples collected at the stockpiles (S-samples).

Once the clayey soil was mined and stockpiled, a second set of samples were collected from the stockpiles. They were designated as S-samples to indicate the method of collection. During stockpiling, accurate records were kept so that the original location and the depth of the soil could be easily identified to compare S-samples with previously collected samples D-samples. Depth and location identification was conducted by utilizing a hand held GPS unit. When the area of the clay mine was excavated, the approximate location of this material was flagged in the stockpiles. By doing this, it was ensured that the S-soil samples were also from the same locations where the D-samples were collected. S-samples were collected and transported in five gallon buckets.

About 65 S-samples were finally collected to compare with D-samples. Samples retrieved from this method did not have a fine consistency as the D-samples, and instead the S-samples contained many clay clods, as seen in Figure 1(b).

2.2. Sample Processing

For the initial soil characterization, soil samples were processed either completely manually or using a soil processor. During manual processing, soil samples were first allowed to air dry at room temperature then placed over a #4 screen and pushed through by hand. This manual preparation was designated as M-processing.

During sample preparation using the soil processor, air dried samples were first placed in the soil processor. Then the weighted lid of this machine applied a normal dead load to the sample as an air-operated arm moves the #4 screen perpendicular to the sample. The results produced by this method were more or less identical to the soil samples produced by the above process and hence this preparation method was also designated as M-processing.

D-samples did not need much of a preparation other than air drying as they were already fine in consistency. However, to make the research process consistent, D-samples were also nominally processed manually by sieving through the #4 screen. Soil clods in S-samples needed to be processed before testing. An example of a manually processed S-sample is shown in Figure 2(a). The #4 screen used in M-processing can be seen in Figure 2(b).

(a)

(b)

Figure 2. (**a**) A stockpile sample processed manually using #4 screen (SM) (**b**) #4 screen used in manual processing.

2.3. Geotechnical Characterization

ASTM test protocols were followed during geotechnical testing. To characterize the material, particle size analysis (ASTM D421 and D422), Atterberg limits (ASTM D4318), classification (ASTM D2487), modified Proctor (ASTM D1557) and hydraulic conductivity (ASTM D5084) tests were carried out as per the ASTM test protocols [4]. Hydraulic conductivity tests were slightly modified to suit the requirements of this research. Before testing for hydraulic conductivity, the samples were placed in a sealed container to temper for at least 16 hours. Then samples were remolded (mold dimensions: 2.79″ diameter × 3″ height) and compacted to 90 percent of the maximum dry unit weight, as determined by the modified Proctor test. Remolded specimen was then placed within a flexible wall permeameter cell.

3. RESULTS, ANALYSIS AND DISCUSSION

All D-samples and S-samples were first processed manually (hence designated as DM and SM) and subjected to mechanical analysis to obtain their gradation.

Size distributions of all theses samples are given in Moran [5]. Based on the results of mechanical analysis, 13 DM-samples with similar clay content (approximately 28%) were selected for further analysis. Size distributions of these 13 DM-samples are given in Table 1. SM-samples, those were originated from the same locations as in DM-samples were also selected for further processing. Size distributions of the 13 SM-samples are given in Table 2. As is evident from Tables 1 and 2, the grain size distributions of DM-samples do not match with the grain size distributions for SM-samples, though they originated from the same locations. The difference in clay content was nearly 10 percent. To illustrate this difference in clay content, grain-size distributions for DM-1 and SM-1 are compared in Figure 3.

Both DM and SM soil samples were then subjected to further geotechnical characterization through Atterberg limits, classification per USCS, modified Proctor and hydraulic conductivity tests. According to the results of these tests summarized in Table 1 and Table 2, the material is rather homogenous. Hydraulic conductivity is typically a function of the clay content. But the 10% difference in the clay content has not even made any considerable impact on hydraulic conductivity. Therefore it is clear that the clay content of one of the sample types is incorrect.

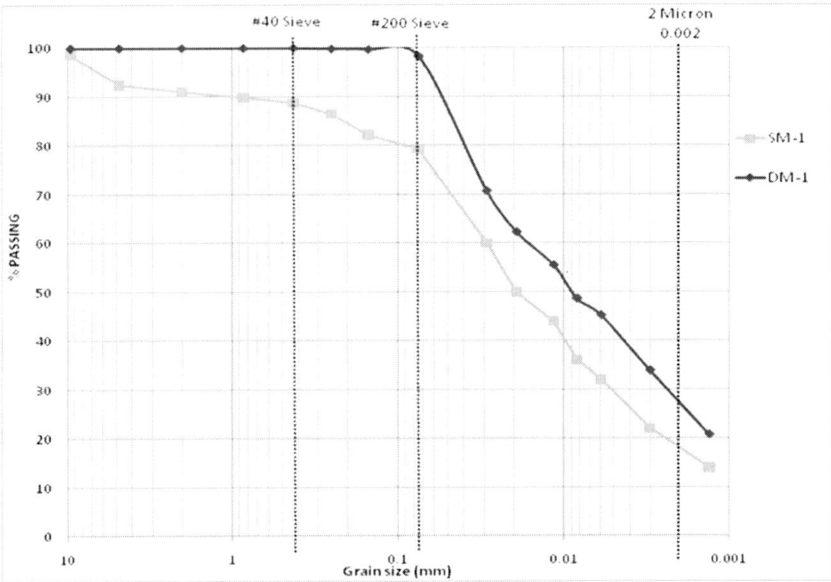

Figure 3. Comparison of grain size distributions: manually processed drilled sample (DM-1) and manually processed sample collected at the stockpile (SM-1).

Table 1. Geotechnical characteristics of DM-samples selected for further processing.

Sample Number	Grain Size Distribution						Atterberg Limits[1]			Uses Soil Type	Compaction Characteristics Based On Modified Proctor Test[2]		Hydraulic Conductivity (cm/sec)
	% Finer 12.5 mm	% Finer 9.5 mm	% Finer #4	% Finer #40	% Finer #200	% Finer .002 mm	LL	PL	PI		γ_{dmax} (kN/m^3)	OMC (%)	
DM-1	100.0	100.0	100.0	100.0	98.3	27.5	30	16	14	CL	–	–	–
DM-2	100.0	100.0	100.0	82.1	69.4	27.7	29	16	13	CL	21.2	6.6	7.6×10^{-8}
DM-3	100.0	100.0	100.0	99.5	93.2	27.5	27	17	10	CL	22.0	7.1	9.0×10^{-8}
DM-4	100.0	100.0	100.0	94.9	83.2	28.3	28	15	13	CL	–	–	–
DM-5	100.0	100.0	100.0	99.7	96.1	28.1	24	14	10	CL	–	–	–
DM-6	100.0	100.0	100.0	98.9	92.6	28.8	23	15	8	CL	–	–	–
DM-7	100.0	100.0	100.0	83.3	73.3	29.8	26	16	10	CL	–	–	–
DM-8	100.0	100.0	100.0	96.5	89.1	27.7	28	16	12	CL	–	–	–
DM-9	100.0	100.0	100.0	98.9	94.6	27.2	28	18	10	CL	–	–	–
DM-10	100.0	100.0	100.0	80.7	74.4	27.7	27	17	10	CL	–	–	–
DM-11	100.0	100.0	100.0	98.9	95.9	27.6	33	19	14	CL	–	–	–
DM-12	100.0	100.0	100.0	99.8	95.4	28.2	25	17	8	CL	–	–	–
DM-13	100.0	100.0	100.0	99.8	93.8	27.6	26	17	9	CL	–	–	–

[1] LL = liquid limit, PL = plastic limit, PI = plasticity index;

[2] γ_{dmax} = maximum dry unit weight, OMC = optimum moisture content.

Table 2. Geotechnical characteristics of SM-samples selected for further processing.

Sample Number	Grain Size Distribution						Atterberg Limits[1]			Uses Soil Type	Compaction Characteristics Based On Modified Proctor Test[2]		Hydraulic Conductivity (cm/sec)
	% Finer 12.5 mm	% Finer 9.5 mm	% Finer #4	% Finer #40	% Finer #200	% Finer .002 mm	LL	PL	PI		γ_{dmax} (kN/m³)	OMC (%)	
SM-1	100.0	98.5	92.4	88.7	79.3	17.3	30	17	13	CL	21.9	7.3	8.5×10^{-8}
SM-2	100.0	100.0	94.6	87.8	75.7	17.4	28	18	10	CL	21.3	5.9	7.6×10^{-8}
SM-3	100.0	100.0	95.5	87.1	76.9	17.2	28	18	10	CL	21.9	8.0	9.1×10^{-8}
SM-4	100.0	100.0	92.5	91.8	83.1	18.2	29	17	12	CL	21.7	7.0	6.5×10^{-8}
SM-5	100.0	100.0	98.7	94.6	85.6	17.3	29	18	11	CL	–	–	–
SM-6	100.0	99.4	96.0	89.0	78.1	18.5	24	16	8	CL	–	–	–
SM-7	100.0	100.0	99.5	93.9	83.7	20.6	27	17	10	CL	–	–	–
SM-8	100.0	99.2	94.1	89.2	79.0	18.8	28	16	12	CL	–	–	–
SM-9	100.0	100.0	95.6	93.2	80.2	19.5	28	19	9	CL	–	–	–
SM-10	100.0	99.2	97.5	90.7	78.6	19.3	28	17	11	CL	–	–	–
SM-11	100.0	100.0	100.0	92.8	82.4	19.1	31	19	12	CL	–	–	–
SM-12	100.0	100.0	98.4	88.9	78.1	17.5	25	17	8	CL	–	–	–
SM-13	100.0	99.4	94.5	88.1	78.0	17.6	27	17	10	CL	–	–	–

[1] LL = liquid limit, PL = plastic limit. PI = plasticity index;

[2] γ_{dmax} = maximum dry unit weight, OMC = optimum moisture content.

3.1. Sample Processing by a Soil Grinder

During the next phase of the research, the laboratory processing of S-samples were performed using a soil grinder (hence SG-samples). The mechanical soil grinder utilized is the one typically used for preparing samples for Atterberg limits and specific gravity tests. Air dried samples were placed into the soil grinder which consists of an electric motor driving a ½ inch shaft with ¼ inch spinning steel bars. The grinder "broke the soil down" to clods to the approximate size of a number 10 sieve or smaller. This method was designated as G-processing (hence processed samples designated as SG-samples). An example of a sample prepared by this method is shown in Figure 4(a) and the apparatus for grinding can be seen in Figure 4(b). Visual appearance of the ground samples appeared similar to the samples collected through the reverse circulation drilling exploration (D-samples).

(a)

(b)

Figure 4. (**a**) A stockpiled sample processed through grinding (SG); (**b**) the soil grinder used for processing.

The results of particle size analysis, Atterberg limits, modified Proctor and hydraulic conductivity tests conducted on the SG-samples are displayed in Table 3. Figure 5 compares the grain-size distributions for SM-1 and SG-1 samples (from the same location). The plot shows a clear difference in the clay content in the gradation curves obtained from manual processing and grinder processing of the same S-samples.

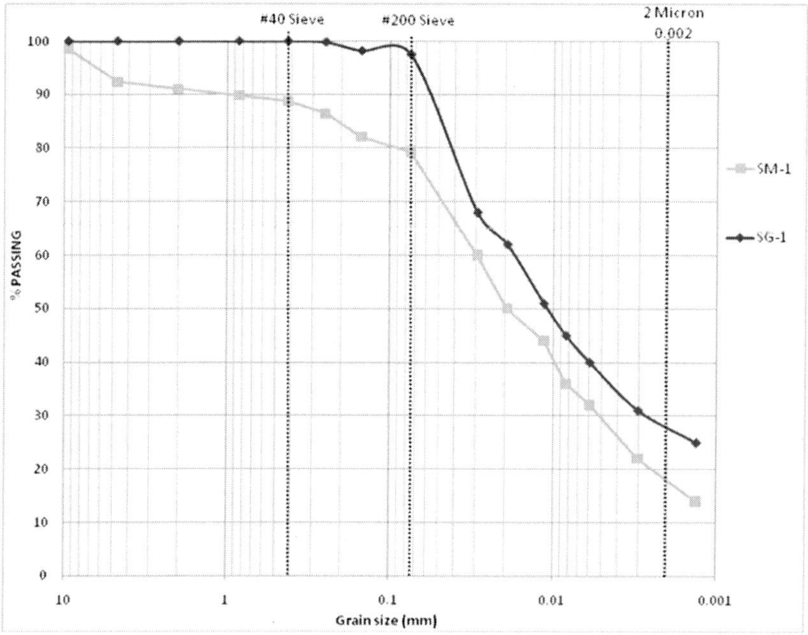

Figure 5. Comparison of grain size distributions: manually processed stockpiled sample (SM-1) *vs.* grinder processed stockpiled sample (SG-1).

However, referring back to Tables 1–3, little to no change is observed in other geotechnical characteristics. The small variations noted are well within the typical expected test margins of error for Atterberg limits, modified Proctor and hydraulic conductivity tests. On the other hand, the grain size distributions of SG-samples (Table 3) nearly mimic the results for the DM-samples (Table 1). Figure 6 compares the grain-size distributions of DM-1 and SG-1 samples.

The results of the laboratory samples processed using a soil grinder, indicate the impact of the mechanical process to have a direct affect on the grain-size distribution. Since the same S-samples showed a clay content similar to that of DM when processed using a grinder (SG), it is clear that the discrepancy could be due to the soil clods that are not easily broken during processing. Standard ASTM test protocols for sieve and hydrometer tests have been in use for some time and they have an excellent track record for producing repeatable and reliable data. The question that needs an answer is that why mined clay samples collected at the stockpiles have become an exception.

Table 3. Geotechnical characteristics of SG-samples.

Sample Number	Grain Size Distribution % Finer 12.5 mm	% Finer 9.5 mm	% Finer #4	% Finer #40	% Finer #200	% Finer .002 mm	Atterberg Limits[1] LL	PL	PI	USCS Soil Type	Compaction Characteristics Based On Modified Proctor Test[2] γ_{dmax} (kN/m³)	OMC (%)	Hydraulic Conductivity (cm/sec)
SG-1	100.0	100.0	100.0	100.0	97.5	28.9	29	15	14	CL	21.9	7.27	8.4×10^{-8}
SG-2	100.0	100.0	100.0	90.0	80.0	20.2	27	19	8	CL	21.0	7.20	8.1×10^{-8}
SG-3	100.0	100.0	100.0	98.4	92.4	29.2	27	18	9	CL	22.0	7.08	8.8×10^{-8}
SG-4	100.0	100.0	100.0	91.8	89.5	28.1	28	18	10	CL	21.4	7.85	6.9×10^{-8}
SG-5	100.0	100.0	100.0	96.5	95.4	29.1	29	18	11	CL	–	–	–
SG-6	100.0	100.0	100.0	98.4	92.0	28.0	25	15	10	CL	–	–	–
SG-7	100.0	100.0	100.0	96.1	88.7	25.0	26	16	10	CL	–	–	–
SG-8	100.0	100.0	100.0	97.2	94.5	27.5	29	15	14	CL	–	–	–
SG-9	100.0	100.0	100.0	96.4	95.2	27.4	29	15	14	CL	–	–	–
SG-10	100.0	100.0	100.0	98.5	94.2	29.5	30	18	12	CL	–	–	–
SG-11	100.0	100.0	100.0	96.5	91.0	26.0	31	20	11	CL	–	–	–
SG-12	100.0	100.0	100.0	91.0	82.5	22.4	26	16	10	CL	–	–	–
SG-13	100.0	100.0	100.0	99.1	96.9	29.2	28	17	11	CL	–	–	–

[1] LL = liquid limit. PL = plastic limit, PI = plasticity index;

[2] γ_{dmax} = maximum dry unit weight, OMC = optimum moisture content.

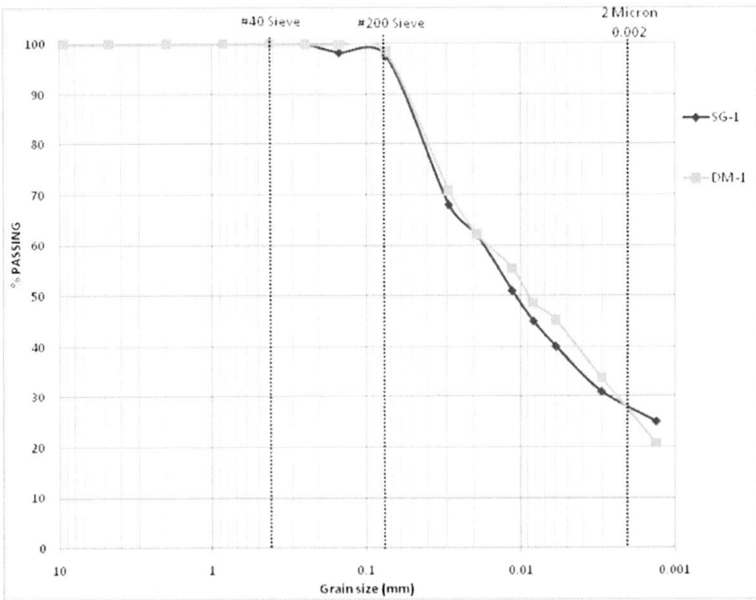

Figure 6. Comparison of grain size distributions: Manually processed drilled *vs.* grinder processed stockpiled samples (DM-1 & SG-1).

3.2. Stockpile Mined Clay Samples from Other Locations in the Midwest

To see if the use of a soil grinder has any impact on the grain size distribution, comparable clayey soil samples obtained from two other locations in the Midwest were analyzed. Four samples from southwestern Michigan and another four from central Kentucky were used for this purpose. Samples were split in half prior to processing. One set of samples was first processed by hand through a number 4 screen. The other set of samples was then processed using the soil grinder after air drying. Both sets were then subjected to gradation analysis, Atterberg limits, modified Proctor, and hydraulic conductivity testing.

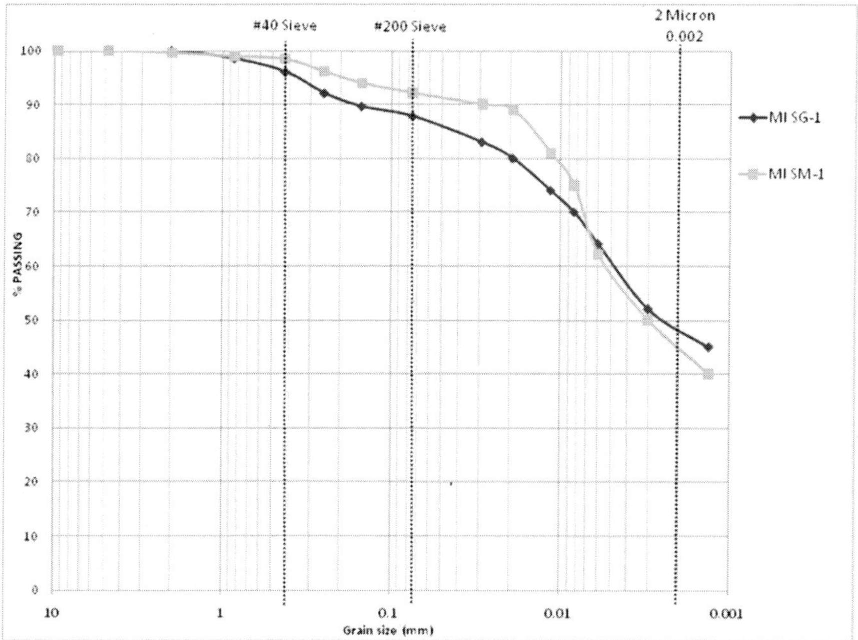

Figure 7. Comparison of grain size distributions: manually processed *vs.* grinder processed stockpiled samples of mined clay from Southwest Michigan (MI SM-1 & MI SG-1).

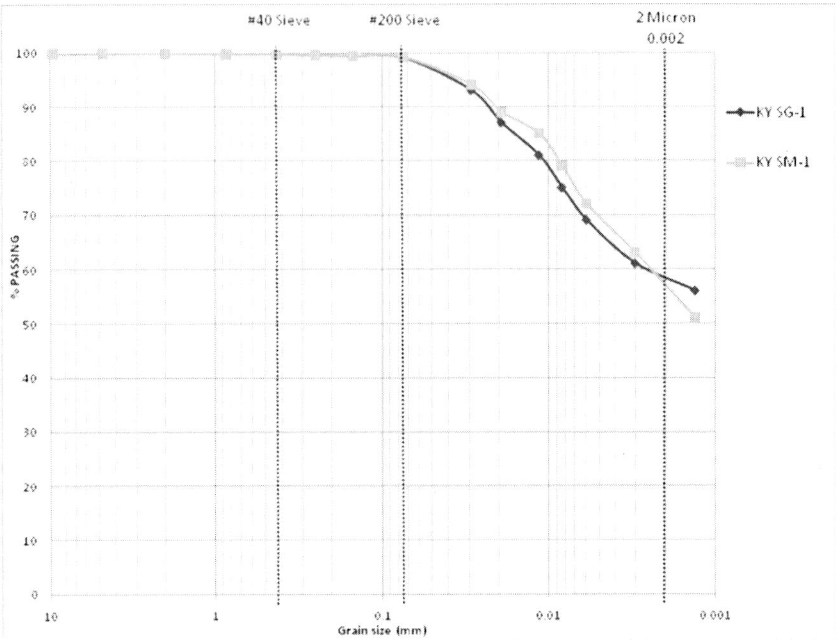

Figure 8. Comparison of grain size distributions: manually processed *vs.* grinder processed stockpiled samples of mined clay from Central Kentucky (KY SM-1 & KY SG-1).

Table 4. Geotechnical characteristics of stockpiled mined clay-samples from Southwest Michigan processed manually and using a soil grinder.

Sample Number	Grain Size Distribution						Atterberg Limits[1]			USCS Soil Type	Compaction Characteristics Based On Modified Proctor Test[2]		Hydraulic Conductivity (cm/sec)
	% Finer 12.5 mm	% Finer 9.5 mm	% Finer #4	% Finer #40	% Finer #200	% Finer .002 mm	LL	PL	PI		γ_{dmax} (kN/m³)	OMC (%)	
MI SM-1	100.0	100.0	99.6	98.4	92.1	45.5	38	24	14	CL	18.8	12.45	6.8×10^{-8}
MI SM-2	100.0	100.0	100.0	99.2	93.9	31.9	34	16	18	CL	19.1	12.36	5.4×10^{-8}
MI SM-3	97.0	96.5	94.2	86.5	62.8	21.1	24	16	8	CL	–	–	–
MI SM-4	100.0	100.0	99.0	98.2	96.7	50.7	39	24	15	CL	–	–	–
MI SG-1	100.0	100.0	100.0	96.0	87.8	48.9	38	24	14	CL	18.6	12.55	7.0×10^{-8}
MI SG-2	100.0	100.0	100.0	100.0	96.5	38.2	35	16	19	CL	19.0	12.44	5.3×10^{-8}
MI SG-3	100.0	100.0	98.5	88.2	63.5	25.2	27	16	11	CL	–	–	–
MI SG-4	100.0	100.0	99.2	98.2	96.9	51.0	40	22	18	CL	–	–	–

[1] LL = liquid limit, PL = plastic limit, PI = plasticity index;

[2] γ_{dmax} = maximum dry unit weight, OMC = optimum moisture content.

Table 5. Geotechnical characteristics of stockpiled mined clay-samples from Central Kentucky processed manually and using a soil grinder.

Sample Number	Grain Size Distribution						Atterberg limits[1]			USCS Soil Type	Compaction Characteristics Based On Modified Proctor Test[2]		Hydraulic Conductivity (cm/sec)
	% Finer 12.5 mm	% Finer 9.5 mm	% Finer #4	% Finer #40	% Finer #200	% Finer .002 mm	LL	PL	PI		γ_{dmax} (kN/m^3)	OMC (%)	
KY SM-1	100.0	100.0	99.9	99.7	99.2	58.6	47	21	26	CL	18.0	15.27	3.2×10^{-8}
KY SM-2	100.0	100.0	99.9	99.8	99.2	42.5	50	22	28	CH	18.5	13.89	4.1×10^{-8}
KY SM-3	100.0	100.0	99.3	97.9	96.6	54.6	53	24	29	CH	–	–	–
KY SM-4	100.0	100.0	100.0	99.8	99.4	50.9	52	27	25	CH	–	–	–
KY GR-1	100.0	100.0	100.0	99.8	99.1	58.0	46	21	25	CL	18.3	14.91	3.4×10^{-8}
KY GR-2	100.0	100.0	99.8	99.6	99.0	43.0	51	24	27	CH	18.5	14.01	4.1×10^{-8}
KY GR-3	100.0	100.0	99.8	98.0	96.8	54.2	54	25	29	CH	–	–	–
KY GR-4	100.0	100.0	100.0	99.9	99.3	50.1	50	25	25	CH	–	–	–

[1] LL = liquid limit, PL = plastic limit, PI = plasticity index;

[2] γ_{dmax} = maximum dry unit weight, OMC = optimum moisture content.

Included in Table 4 and Table 5 are the results from the tests conducted on samples from southwestern Michigan and central Kentucky respectively. No significant change is noted in the Atterberg limits, modified Proctor, hydraulic conductivity as well as the gradation of the test specimen based on the processing method. Figure 7 compares the gradation curves of the mined clay from southwestern Michigan collected at the stockpiles but processed manually (MI SM-1) and using a soil grinder (MI SG-1). Figure 8 presents the same comparison for the samples from Central Kentucky (KY SM-1 and KY SG-1). Figure 7 and Figure 8 provide examples of how well the gradation curves from two methods compare with each other. The conclusion these results provide is that the discrepancy observed in mined clay from Appalachian Ohio is an exception and may not be due to a deficiency of the standard testing methods.

3.3. Geological Background of Clay Deposits in Appalachian Basin

Information on the regional geology and the composition of mined clay from Appalachian Ohio was taken into consideration to investigate the discrepancies it shown in the grain size analysis. As explained before, soil found beneath coal layers at the base of the Appalachian basin is comprised of shale enriched clay deposits [1]. Shale in this region is mostly of lightly indurated silt and clay that can be broken with little force applied with the addition of moisture. The shale, depending on moisture content, will either appear as a gravel sized particle or as a clay clod. With the cyclothemic affects in Appalachian Ohio, one would expect a large variety and discontinuity within and between the sedimentary layers. Natural moisture content makes a direct influence on the visual classification of this soil. The higher the natural moisture content, the more likely the material will be visually classified as clayey soil and not as shale.

This specific composition of mined clay from Appalachian Ohio reveals that the soil clods observed in the stockpile samples are mostly the fragments of shale. Manual processing does not break all shale clods. The dispersing agent used during the hydrometer test is only able to separate some of the fines into individual particles, but not strong enough to break the shale fragments into the respective parent materials (*i.e.* clay and silt). Therefore, it is not surprising to see a lower clay fraction in SM-samples, because the small shale fragments may have behaved similar to coarse grained soils during the mechanical analysis. When a grinder was used to process soils, it was able to break the shale fragments into smaller clods. These smaller clods must have allowed the dispersing agent to more readily separate the shale fragments into its parent particles and increase the clay fraction in SG-samples. This explains why SG-samples behaved similar to DM-samples during mechanical analysis (Figure 6).

On the other hand, the mined clay from southwestern Michigan was from a glacial deposit and samples from central Kentucky were from a cyclothemic deposit. These samples from Michigan and Kentucky were high in clay content with traces of sand. Unlike the soil samples from Appalachian Ohio, none of these other samples would be considered to contain any shale fragments. As a result, they did not exhibit a considerable difference in the grain size distributions based on the processing method. Therefore, the presence of shale fragments should be the primary cause for the discrepancies observed in the grain-size analysis.

3.4. Impact on Clay Mining Industry

Hydraulic conductivity is the most important characteristic of RSL material and The State of Ohio has used 20 percent as a requirement for borrow soils used in RSL systems [2]. As per the results shown in Table 1 through Table 5, SM samples do not meet the 20% clay content requirement, but SG and DM-samples do meet this requirement. However, it is interesting to note that all three types of samples are from the same origin and they share the same hydraulic conductivity. Hydraulic conductivity values obtained for all three types of samples are compared in Figure 9. As can be seen in Figure 9 all hydraulic conductivity values are bordering along 1.0×10^{-7} cm/sec regardless of the sample collection/processing method.

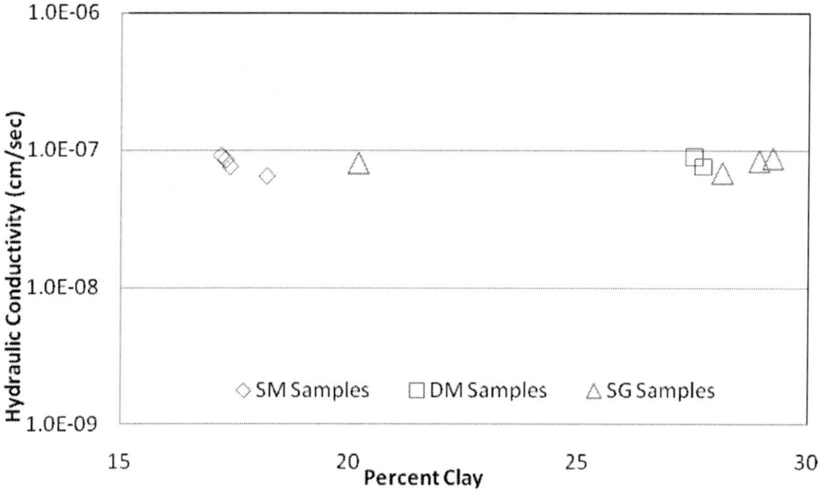

Figure 9. Hydraulic conductivity *vs.* percent clay content obtained for DM, SM and SG samples.

When soil samples are delivered to a geotechnical laboratory for testing, geological history, sampling methods, and the expectations of soil evaluation results are seldom provided to the laboratory staff. The geotechnical laboratory is typically expected to act as an independent third party in the soil evaluation process even if the laboratory is within the same company as the sampling technician. This independent behavior works well when the testing standards prescribed within the industry include limited error margins. However, this intentional lack of communication between the geotechnical laboratory and the geotechnical engineer can cause problems when soils with unique characteristics such as the mined clay from Appalachian Ohio are tested.

If not instructed specifically, a laboratory may or may not use a grinder for processing soils, and depending on that decision, even a suitable soil may very well be rejected. It was reported that in 2003, over 2.1 million tons of clay were mined in Ohio [6]. With prices averaging $20/ton, selling mined clay for RSL construction is significant to Ohio's mining industry. Therefore, decision making purely based on the apparent clay content without paying much attention to the hydraulic conductivity, can make a considerable negative impact on the clay mining industry.

4. CONCLUSIONS

The following conclusions were from the material discussed in this research. It should be noted that all conclusions are exclusively applicable to mined clay from Appalachian Ohio and may not be applicable to other general type of clayey soils.

a. Grain size distributions of mined clay from Appalachian Ohio exhibit sensitivity to sampling and/or laboratory processing methods. Other geotechnical characteristics such as Atterberg limits, compaction (based on modified Proctor method) and hydraulic conductivity were not sensitive to sampling and or laboratory processing methods.

b. The Presence of shale fragments in mine clay from Appalachian Ohio was found to be the reason for observing sampling and laboratory processing dependant results in particle size analysis of mined clay from Appalachian Ohio.

c. Manual processing is not sufficient to separate shale fragments into individual silt and clay particles in the soil samples collected at mined clay stockpiles. Samples collected through drilling or samples collected at the stockpiles but processed using a soil grinder, seem to be producing more reliable grain size distribution data for mined clay from Appalachian Ohio.

d. Use of clay content as the sole criteria to accept or reject Mined clay from Appalachian Ohio for RSL applications could produce misleading results. If clay content is the only criteria to be used, the samples for mechanical analysis should be collected through drilling or else samples collected at stockpiles should be processed using soil grinder.

REFERENCES

1. Coogan, AH. Ohio Surface Rocks and Sediments. In *Fossils of Ohio*; Rodney, MF, Ed.; Ohio Geological Survey: Columbus, OH, USA, 1996; Bulletin 70; Volume Chapter 3, pp. 31–50.
2. *Sanitary Landfill Facility Construction, OAC-3745-27-08*; Ohio Environmental Protection Agency: Columbus, OH, USA, 2003.
3. Das, BM. *Principles of Geotechnical Engineering*, 6th ed; Thomson Canada Limited: Toronto, Ontario, Canada, 2006.

4. *ASTM Annual Book of Standards, Section Four, Construction. Soil and Rock (i)*; ASTM International: West Conshohocken, PA, USA, 2008; pp. D420–5876.

5. Moran, A. Characterization of Mined Clay from Appalachian Ohio: Implications of Sampling and Laboratory Processing Techniques on Local Regulations. M.D. thesis, Department of Civil Engineering, Lawrence Technological University, Southfield, MI, USA, 2009.

6. Wolfe, M. *Rocks and Minerals Mined in Ohio and Their Uses, GeoFacts No. 11*; The Division of Geological Survey, Ohio Department of Natural Resources: Columbus, OH, USA, 2005.

10

CHAPTER

STRUCTURAL DEFORMATION MONITORING AND ANALYSIS OF HIGHWAY BRIDGE USING ACCURATE GEODETIC TECHNIQUES

Ashraf Abd El-Wanis Beshr

Public Works Department, Faculty of Engineering, Mansoura University, Mansoura City, Egypt

ABSTRACT

As the new materials and technologies are increasingly applied to construction of civil infrastructures such bridges, dam and tunnels, the need for structural monitoring systems, maintenance and restoration becomes more important and vital. Bridges are widespread in every society and affect its human, social, economical and cultural aspects. Measurements and monitoring of the structural deformation of highway bridges have an essential role in structural safety. This paper investigates an integrated monitoring system for estimation of the deformation behavior of one of the important reinforced bridges in Egypt. The applied data for analysing the deformation of any structure from geodetic observations are the coordinates of several monitoring points distributed on the structure itself. The coordinates of these points are calculated with respect to control fixed points. So any deviations in the control points coordinates between the two successive epochs of observations will affect the values of structural deformation. To overcome this shortage, applying the multi-parameter transformation will be studied in this paper for structural health monitoring of bridges. Statistical tests using F-Fisher criterion with a confidence level of 98% of the geodetic observations for bridge deformation values are also presented. The results of the practical measurements, analysis of the interesting deformation technique and traffic flow around the studied highway bridge are also presented. The resulting structural deformation values from statistical tests provide a significant improvement of understanding and prediction the structure deformation values of highway bridges.

Keywords: Bridges, Monitoring, Statistical Tests, Transformation, Deformation

1. INTRODUCTION

Structural health monitoring is the term that describes the range of systems implemented on various civil infrastructures, which inform operators whether the structural deformation is the same as those the structure was designed to tolerate or not [1] . The demand for structural health monitoring systems for bridges has grown over the last few decades [2] . In order to know the safeties of bridges, monitoring their real-time displacement and recording their fatigue history are very important.

The measuring techniques and instrumentation for geometrical monitoring of structural deformations have traditionally been categorized into two groups according to the two main groups of professionals who use the techniques [3] :

1) Geodetic surveying, which includes Global Position System (GPS), close range Photogrammetry, terrestrial surveying (leveling, theodolite, total station and very long base line interferometer) and satellite laser ranging.

2) Geotechnical structural measurements of local deformations using tilt meters, strain meters, extensometers, joint meters, laser distance gauge method, etc.

Each type of the measurements has its own advantages and drawbacks [1] . The acceleration integration method integrates the acceleration, which is measured by acceleration gauge, to obtain the displacement. But its error is relatively large. The laser distance gauge method is often influenced by the weather. Furthermore, the geotechnical structural measurements often need to stop the traffic, which brings a lot of costs. So these methods are suitable for some structures whose survey distance is relatively short and the displacement is relatively small. But to structures such bridges, these methods are difficult to use.

Geodetic surveys, through a network of points interconnected by angles and distances measurements, usually supply a sufficient redundancy of observations, for the statistical evaluation of their quality and for detection of errors [4] [5] . They give global information on the behavior of the deformable object. Recent advances in technology have made it possible to use accurate surveying equipments to measure and monitor the deformation of structural members. The emergence of reflectorless accurate total station allows working without special reflectors (prisms). It is now possible to measure without long and tedious search of prisms to lift the reflector under the roof of buildings. The principle of work of reflectorless total station is the same as that of a simple total station [6] .

The surveillance of an object involved in a deformation process requires the object as well as the process to be modeled. Conventionally, geodetic modeling the object (and its surrounding) means dissecting the continuum by discrete points in such a way that the points characterize the object, and that the movements of the points represent the movements and distortions of the object. This means that (only) the geometry of the object is modeled [6] . Furthermore, modeling the deformation process means conventionally to observe the characteristic points in certain time intervals in order to monitor properly the temporal course of the movements. This means that (only) the temporal aspect of the process is modeled. Nowadays, for the analysis and the interpretation of structural deformations, different deformation models have been developed [7] . In deformation monitoring studies, these models include static, kinematic and dynamic models. Static model that is not dependent on time provides the determination of deformations on the characteristic points of the area or the structure, which is monitored. In this paper, statistical tests of geodetic observations using F-Fisher criterion with a confidence level of 98% were applied. Kinematic models allow estimating the velocity and even the acceleration (by building double differences) of monitoring point movements. The intention of kinematic models is to find a suitable description of point movements by time functions without regarding the potential relationship to causative forces.

In Egypt, there are many highway bridges with different lengths and widths which have a great effect on Egyptian life and economy. These bridges were constructed from different materials such as reinforced concrete, steel and wood. In this paper, studying the deformation behavior of one of the most important bridge in Delta Egypt is presented.

2. OBSERVATION SITES AND THE PROPOSED GEODETIC MONITORING TECHNIQUE

Kafr El-Sheikh highway bridge has been chosen for the purpose of the deformation monitoring study. This bridge was suited in Kafr El-Sheikh city. This city is the capital of Kafr El-Sheikh governorate in Egypt and located in the middle of the Nile Delta, where it overlooks the Mediterranean Sea Shore in the north and the River Nile (Rashid Branch) in the west. The international coastal highway (which joins all the Mediterranean Sea countries) passes in the north of it. Kafr El-Sheikh city lies 140 km north of the capital Cairo, 90 km east of Alexandria. The area of Kafr El-Sheikh city is 4.64 km^2. The studied highway bridge connects Kafr El-Sheikh city to the international coastal highway and surrounding cities and villages. Kafr El-Sheikh highway bridge (Figure 1), has a total length 216 m and a width 17 m with average height 7.55 m above ground level. The highway bridge has four lanes in two ways directions. It was made of reinforcement concrete except one span was made of steel (this steel span is constructed over Meet Yazid Channel, Figure 1). The bridge was supported on ten pillars and two abutments. The bridge is subjected to two types of loads; the first type is static load (dead load), which includes the own weight of the bridge (superstructure and covering materials). The second type is moving loads (live loads); these moving loads are the transportation on the bridge.

The suggested monitoring system of bridge consists of small horizontal geodetic network around the bridge and selected monitoring points distributed on the bridge to cover its full length in its two sides. The main objective of the designed network is to provide a number of stations whose relative and absolute positions, are accurately established. More detailed location of monitoring points on the bridge are then carried out from these stations. This network (Figure 2) consists of two braced quadrilaterals figures (each figure contains four corner stations and observed diagonals). For base line observations, two GPS dual frequency receivers were used. The observations of horizontal network were horizontal and vertical angles and 10 distances, such a combined system of measuring angles and distances represents the strongest network for creating horizontal control. The angles were measured on five arcs and the distances were measured 10 times. Then the coordinates of network stations were calculated using least squares adjustment technique. The observations of geodetic network were done only one time at the first epoch of monitoring observations (December

2011). All stations of geodetic networks were fixed in stable positions. Before starting the process of first observations epoch for the bridge monitoring, site reconnaissance and topographic surveying of the bridge area were done.

Each side of the studied bridge contained 23 monitoring points (targets); the spatial distribution of these points should provide complete coverage of the bridge. The monitoring points were located where the maximum deformation values have been predicted (at middle distance of each span), plus a few points which was depending on previous experience could signal any potential unpredictable behavior (targets above each support). These

Positions 1, 2, 3 and 4—Traffic flow count and noise measurementsPositions 2 and 4 are fixed on the bridge

Figure 1. Topographic survey of the studied highway bridge (Kafr El-Sheikh Bridge) and positions of traffic flow count and noise measurements.

CP1, CP2, CP3, ..., CP6—Control Points (stations of geodetic network)RS1, RS2, RS3, RS4—Reference Stations around the highway bridge

Figure 2. The geodetic monitoring network around the studied highway bridge (Kafr El-Sheikh Bridge—Egypt).

points (targets) were located using sheet prisms with diameter (1 cm) fixed on the superstructure of the bridge, which were securely fastened and arranged to be visible from the positions of the used two total stations. Common sense helped to reduce errors in the field. There were four reference stations outside the bridge (Figure 2) and not far from it (RS1, RS2, RS3 and RS4). These stations fixed in stable soil. Using even more reference stations was always a possibility; however, there has never been a discrepancy between all four reference stations during testing to date. The monitoring points on the bridge were observed from the four occupied stations (CP3, CP4, CP5 and CP6; Figure 2). Four epochs of observations for studying the structural deformation of studied bridge were performed at December 2011, May 2012 December 2012 and June 2013.

The used equipments consisted of: Two GPS dual frequency receivers (Sokkia), two total stations GeoMax with accessories which have the accuracy of measuring angles equals to 1'' but the accuracy of measuring distance was 2 mm ± 2 ppm for reflector measuring and 3 mm ± 3 ppm for reflector-less measuring; four tripods, 2 plumb bobs; sheet prisms; two prisms of diameter 5 cm; two survey umbrellas; and two foot plates.

Calibrations of the used instruments were preformed to confirm the specified accuracy for measuring distances and angles. The higher accuracy required for structural health monitoring demanded extra measures to assure desirable results. Most of these additional measures were fairly obvious; the goal was to reduce or eliminate the possibility of undesired instrument and target movement.

3. DETERMINING THE COORDINATES OF MONITORING POINTS ON THE BRIDGE

The two total stations technique was used to determine the spatial coordinates of monitoring points on the bridge and reference stations. This model employees the intersection process in three dimensions coordinates. If there were two known coordinates stations (X_A, Y_A, Z_A) and (X_C, Y_C, Z_C) from geodetic network and there were three unknown parameters (X_B, Y_B, Z_B) and six observations from the two total stations (two slope distances S_1, S_2, two horizontal angles α_1, α_2; and two vertical angles γ_1, γ_2). The parametric least squares adjustment technique will be applied to get the adjusted values of unknowns as following:

The two measured slope distances (S_1, S_2) from total station to the monitoring point on the bridge can be written as:

$$\left.\begin{aligned} S_1 &= \sqrt{(X_B - X_A)^2 + (Y_B - Y_A)^2 + (Z_B - Z_A)^2} \\ S_2 &= \sqrt{(X_B - X_C)^2 + (Y_B - Y_C)^2 + (Z_B - Z_C)^2} \end{aligned}\right\} \quad (1)$$

By using the coordinates formulae, the horizontal angles (α_1 and α_2) can be written in the following form:

$$\left.\begin{aligned} \alpha_1 &= \cos^{-1}\left[\frac{(X_B - X_A)^2 + (Y_B - Y_A)^2 + AC^2 - (X_B - X_C)^2 - (Y_B - Y_C)^2}{2AC\sqrt{(X_B - X_A)^2 + (Y_B - Y_A)^2}}\right] \\ \alpha_2 &= \cos^{-1}\left[\frac{(X_B - X_C)^2 + (Y_B - Y_C)^2 + AC^2 - (X_B - X_A)^2 - (Y_B - Y_A)^2}{2AC\sqrt{(X_B - X_C)^2 + (Y_B - Y_C)^2}}\right] \end{aligned}\right\} \quad (2)$$

The two vertical angles (γ_1 and γ_2) can be calculated in the coordinates form as following:

$$\left.\begin{aligned}\gamma_1 &= \tan^{-1}\left[\frac{Z_B - Z_A}{\sqrt{(X_B - X_A)^2 + (Y_B - Y_A)^2}}\right] \\ \gamma_2 &= \tan^{-1}\left[\frac{Z_B - Z_C}{\sqrt{(X_B - X_C)^2 + (Y_B - Y_C)^2}}\right]\end{aligned}\right\} \tag{3}$$

The Equations (1), (2) and (3) are the six observational equations, these equations are nonlinear function of both parameters and observations; the cofactor matrix ($Q = W^{-1}$) of the observations were formed for each point. This matrix has the dimensions (6, 6) and has the form:

$$Q_{(6,6)} = W_{(6,6)}^{-1} = diag\left[\sigma_{S_1}^2, \sigma_{S_2}^2, \sigma_{\alpha_1}^2, \sigma_{\alpha_2}^2, \sigma_{\gamma_1}^2, \sigma_{\gamma_2}^2\right] \tag{4}$$

In which $\sigma_{S_1}^2$, $\sigma_{S_2}^2$, $\sigma_{\alpha_1}^2$, $\sigma_{\alpha_2}^2$, $\sigma_{\gamma_1}^2$, $\sigma_{\gamma_2}^2$ are the standard deviation of observations and can be determined by instrument calibration. In this model, all observations were assumed to be independent and uncorrelated.

4. APPLYING SEVEN-PARAMETER TRANSFORMATION FOR STRUCTURAL DEFORMATION MONITORING

The used data for analysis the structural deformation of the studied bridge are the coordinates of monitoring points (targets). The coordinates of these points were calculated depending on the occupied stations coordinates. So any deviations in coordinates of control points between original and repeated observations epochs will affect the coordinates of monitoring points and consequently affect the values of structural deformation.

To overcome this great problem, four reference stations were selected not far the studied bridge and fixed in stable soil. The resulted deviations (differences) of these reference stations coordinates between epochs were very small and didn't mean that these points were moved or deformed. In geodesy, there are no identical coordinates of points between two epochs of observations [2] . The deviation in coordinates resulted from systematic errors in observations (temperature, air pressure, instrumental errors, and observations in different times of day... etc.). Then the following steps were done:

1) Carrying out transformation of reference stations coordinates from any other epoch of observations (second, third, fourth...) to the coordinates of the first epoch of observations by using seven parameters transformation to find these parameters.

2) Depending on the resulting parameters of transformation, the coordinates of all monitoring points of the other epochs of observations (second, third and fourth) were transformed to the system of coordinates of the first epoch of monitoring process. By this way, any small deviations in the monitoring points coordinates were removed.

Seven-parameters transformation is used for relating the original and repeated coordinates of the original points using the following form:

$$\begin{bmatrix} X_J^0 \\ Y_J^0 \\ Z_J^0 \end{bmatrix}_{\text{org}} = \begin{bmatrix} T_X \\ T_Y \\ T_Z \end{bmatrix} + (1+\lambda)\cdot R(\omega_X, \omega_Y, \omega_Z)\cdot \begin{bmatrix} X_J^k \\ Y_J^k \\ Z_J^k \end{bmatrix}_{\text{repeated}}, \qquad (5)$$

where: $\begin{bmatrix} X_J^0 \\ Y_J^0 \\ Z_J^0 \end{bmatrix}_{\text{org}}$ = Original coordinates of reference point J at the first epoch of observations; $\begin{bmatrix} T_X \\ T_Y \\ T_Z \end{bmatrix}$ = Vector of translation components; $R(\omega_X, \omega_Y, \omega_Z)$ = Rotation matrix; $\begin{bmatrix} X_J^k \\ Y_J^k \\ Z_J^k \end{bmatrix}_{\text{repeated}}$ = coordinates of the same points in

the other cycle of observations (repeated coordinates); $(1+\lambda)$ = The scale factor; ω_X, ω_Y, ω_Z = the rotation components.

Since the values of rotation components (ω_X, ω_Y, ω_Z) will be too small, the rotation matrix can be simplified and the Equation (5) can be written as:

$$\begin{bmatrix} X_J^0 \\ Y_J^0 \\ Z_J^0 \end{bmatrix}_{\text{org}} = \begin{bmatrix} T_X \\ T_Y \\ T_Z \end{bmatrix} + (1+\lambda)\cdot\begin{bmatrix} 1 & \omega_Z & -\omega_Y \\ -\omega_Z & 1 & \omega_X \\ \omega_Y & -\omega_X & 1 \end{bmatrix}\cdot\begin{bmatrix} X_J^k \\ Y_J^k \\ Z_J^k \end{bmatrix}_{\text{repeated}}, \tag{6}$$

If the identical points in the two system coordinates (in this case, the reference stations near the bridge) are more than three, the least square theory must be applied to find the seven transformation parameters [8] . The mathematical model for this case (parametric least square) will have the form:

$$\underset{(m,7)}{A}\cdot\underset{(7,1)}{\delta}+\underset{(m,1)}{V}=\underset{(m,1)}{L}, \tag{7}$$

where:

m—The number of equations (in this case m = 3n);

n—The number of identical points (for the studied bridge n = the number of reference stations = 4).

The elements of matrix A can be determined by differentiate the Equation (6) with respect to the parameters of transformation. The steps of solving using parametric least squares method were applied to find the seven transformation parameters. If the coordinates of reference stations are identical, the seven parameters will equal zero.

5. STATISTICAL TESTS OF GEODETIC OBSERVATIONS FOR BRIDGE DEFORMATION VALUES

Statistical tests are widely used to determine if a given quantity (coordinates of points in first epoch observations) is compatible with, or significantly different from, some other quantity. Statistical tests on the residuals would determine which of the above statements is supported by the data. For network adjustments, statistical testing is widely used for data quality assessment. So the statistical tests can be used to detect whether there are any significant coordinates differences between each two epochs of observations.

This model of the observations analysis does not consider the time intervals between the observations nor the factors responsible for the deformation—explicitly. The first step in statistical test is forming the vector of the coordinate differences of monitoring points at two different epochs at initial time (first epoch of observation) and t_k, and its cofactor matrix. Point displacements Δ_J are calculated by differencing the adjusted coordinates of this point J for the most recent survey campaign (k), from the coordinates obtained at initial time as follows:

$$\Delta_J = \begin{bmatrix} X_J^K - X_J^0 \\ Y_J^K - Y_J^0 \\ Z_J^K - Z_J^0 \end{bmatrix} = \begin{bmatrix} \Delta X_J \\ \Delta Y_J \\ \Delta Z_J \end{bmatrix}, \tag{8}$$

where:

X_J^K, Y_J^K, Z_J^K = the adjusted coordinates of monitoring point J at time t_K;

X_J^0, Y_J^0, Z_J^0 = the adjusted coordinates of monitoring point J at first time of observations (initial epoch);

$K = 1, 2, \cdots, m$ (m = number of observations epochs);

$J = 1, 2, \cdots, n$ (n = number of monitoring points on the bridge).

The vector of root-mean-square error (RME) of the coordinate differences can be calculated by using the matrix form as following:

$$m_{\Delta_J} = \begin{bmatrix} \sigma^2_{\Delta X_J} & 0 & 0 \\ 0 & \sigma^2_{\Delta Y_J} & 0 \\ 0 & 0 & \sigma^2_{\Delta Z_J} \end{bmatrix} = \begin{bmatrix} \sigma^2_{X_J^K} + \sigma^2_{X_J^0} & 0 & 0 \\ 0 & \sigma^2_{Y_J^K} + \sigma^2_{Y_J^0} & 0 \\ 0 & 0 & \sigma^2_{Z_J^K} + \sigma^2_{Z_J^0} \end{bmatrix}, \tag{9}$$

where:

where:

$\sigma_{X_J^K}, \sigma_{Y_J^K}, \sigma_{Z_J^K}$ = The accuracy of coordinates of point J at time t_K;

$\sigma_{X_J^K}, \sigma_{Y_J^K}, \sigma_{Z_J^K}$ = The accuracy of coordinates of point J at first epoch of observations.

Statistical test analysis for structural deformation observations can be done using several models and criteria for approval in the study of distribution laws of geodetic measurements. In this paper, F-Fisher criterion with a confidence level of 98% was applied to test the movement of each monitoring point on the bridge statistically. For achieving that goal it is important to calculate the factor T_G in dependence on the resulted values of coordinate displacements vector and vector of its accuracy (RME vector) as following:

$$T_G = \frac{\underset{(1,3)}{\Delta^T} \, \underset{(3,3)}{m_\Delta^{-1}} \, \underset{(3,1)}{\Delta}}{h\mu_0^2}, \tag{10}$$

where:

h = Rank of matrix;

μ_0^2 = generalized variance of the process.

The value of generalized variance of the process can be determined by applying the following formula:

$$\mu_0^2 = \frac{f_K \mu_{0_K}^2 + f_0 \mu_{0_{(0)}}^2}{f_K + f_0}, \tag{11}$$

where:

μ_{0k}^2, $\mu_{0_{(0)}}^2$ = variance factors at time k and at first epoch (initial time);

f_K, f_0 —the degree of freedom at time k and at first epoch.

If the test value T_G is greater than the critical value, i.e. $T_G \geq F(h, f, 1-\alpha)$, (where $(1-\alpha)$—confidence level), then there are statistically significant deformation in the common points.

Statistical test was done for each monitoring point on the studied bridge individually for all observations epochs.

6. TRAFFIC FLOW COUNT ON THE STUDIED BRIDGE

Data collected from long-term monitoring have shown that the quasi-static and dynamic response of a bridge may change, depending on the environmental and operating conditions [9] . Contributors are found to be ambient temperature, wind speed and direction, humidity, solar radiation and vehicular loading. All of these loadings lead not only to a direct response from the structure, but can also affect physical parameters and mechanisms in a bridge. Daily traffic is the main live load which has significant impact on the strength and serviceability of bridges. Bridge/vehicle interactions have been studied since the middle of 20th century [10] . The impact of a vehicle on a bridge was initially assumed to be a moving load without considering the inertia effect [11] . Later, a vehicle was simplified as a moving mass which can consider the inertia effect on bridges. In recent years, the commonly-used analytical model for vehicle is a dynamic system consisting of mass, spring and damping parts, which have significant effects on dynamic analysis of vehicles and interactions with other systems. Thus, the interaction analysis of bridge and vehicles is to investigate the coupling nature of multiple dynamic systems. By numerically solving the coupled equations in the time domain, the dynamic response (displacement, acceleration and stress) of the bridge and vehicles can be obtained, respectively [12] . In this paper, traffic flow count and noise measurements were preformed in the two directions of the studied bridge (Figure 1).

The evolution of traffic flow is complicated in terms of vehicle number, vehicle type combination and drivers' behavior such as lane-changing, acceleration or deceleration [13] . With relation to the volume of traffic using the road, the passenger car is adopted as the standard unit and other vehicles are assessed in terms of passenger car units (pcu). Average daily flow from traffic count in the first two epochs of observations are presented in Table 1 for the selected four count positions (Figure 1). Simple regression analysis was done to indicate the relationship between time and traffic volume.

Noise measurements were carried out in the selected positions near the bridge using 407730 Digital Sound Level Meter. The device was used after calibration by using a calibration unit of model. The Digital Sound Level Meter measures and displays sound pressure levels in dB from 40 to 130 dB. The duration of measuring was 5 hours (from 9:00 am to 2.00 pm) and the readings were recorded every 10 minutes. All readings were recorded when the digital sound level meter is placed at distance of 0.5 m from the body of the observer and at height of 1.20 m from the floor level. From the results of traffic flow count and noise measurements, it is deduced that:

- About 40% of all traffic flow cross on the studied bridge;
- All the recorded noise levels in all positions around the studied bridge are much more the permissible limits according to the Egyptian law of environmental and world organizations limits;
- The noise level decreases when the traffic volume increases;
- Simple regression analysis indicated that, traffic volume and running speed increases the noise level. Whereas traffic volume only decreases the noise level in undivided roads.

Table 1. The values of average traffic flow count around the studied highway bridge.

Day Time	Average traffic flow count (pcu) Average values of days 9, 10 and 11 December 2011				Average traffic flow count (pcu) A verage values of days 15, 16 and 17 May 2012			
	Pos. 1	Pos. 2	Pos. 3	Pos. 4	Pos. 1	Pos. 2	Pos. 3	Pos. 4
7.00 am - 8.00 am	278	194	243	175	281	99	210	119
8.00 am - 9.00 am	329	263	310	273	318	163	254	139
9.00 am - 10.00 am	311	251	352	264	342	162	245	175
10.00 am - 11.00 am	375	281	561	184	452	254	193	105
11.00 am - 12.00 am	245	236	573	200	385	144	205	126
12.00 am - 1.00 pm	640	251	282	258	630	175	192	177
1.00 pm - 2.00 pm	803	362	353	281	538	224	266	114
2.00 pm - 3.00 pm	459	232	345	302	330	181	244	220
Total count (pcu)	3440	2070	3019	1937	3276	1402	1809	1175
Average traffic flow Pcu/hour	430	259	377	242	410	175	226	147

7. RESULTS AND ANALYSIS

The steps for bridge monitoring were carried out according to the following procedure:

1) In the first epoch of observations (December 2011), the coordinates of the first two control points (CP1 and CP2, Figure 2) were determined from GPS observations. Then the horizontal geodetic network were fixed on stable positions and observed (16 horizontal, vertical angles

and 7 distances). The most probable values were then calculated for all angles and distances.

2) The adjusted coordinates of all stations of network were then determined depending on the coordinates of the two control point (CP1 and CP2, Figure 2) using conditional least squares adjustment technique.

3) The reference stations and monitoring points (targets) for each bridge side were observed (13 points on each side) from two total stations fixed at control points in the same time. Then the coordinates of all points were calculated applying the mathematical model presented for each epoch of observations.

4) Before determining the values of deformation, the coordinates of reference stations must be checked to have the same coordinates. In our case there were small deviations between these coordinate as mentioned in Table 2. So Seven parameters transformation were applied to remove these deviations for all observation epochs.

5) Depending on the resulted new coordinates of monitoring points after applying transformation parameters, the analysis (statistical test) of observations were done.

By applying the least square adjustment technique, The adjusted coordinates and its associated accuracy of each point in the monitoring network on the bridge were calculated for all epochs. The standard deviations of monitoring points coordinates for the first observations epoch on the bridge (December 2011) were varied from 1.26 mm to 4.38 mm in the horizontal components, and from 1.3 mm to 7.63 mm in vertical component. While in the second epoch (May 2012), the corresponding values are varied from 1.19 mm to 4.93 mm in horizontal components and from 1.52 mm to 7.59 mm in vertical component. For third observation epoch (December 2012), the standard deviations points coordinates were varied from 0.88 mm to 5.24 mm in the horizontal components, and from 0.79 mm to 6.77 mm in vertical component. Practically, the overall data analysis has shown that the qualities of the observation for all campaigns are good.

The deviation of reference stations coordinates between four observations epochs and the calculated seven parameters transformation were done. Sample of results is presented in Table 2. From the results, it is deduced that the deviations in reference station coordinates were very small for all observations epochs and varied from 42.23 mm to 1.19 mm.

Table 2. The deviation of reference stations coordinates between three observations epochs and the calculated seven parameters transformation.

Reference station		Difference of reference stations coordinates between epochs, mm		Resulted Seven Parameters of transformation		
		2nd epoch - 1st epoch	3rd epoch - 1st epoch	Parameter	2nd epoch - 1st epoch	3rd epoch - 1st epoch
	X	25.10	17.31	T_X, m	0.18861	0.1534
RS1	Y	−29.49	17.50	T_Y, m	0.0981	0.1224
	Z	−1.19	−4.10	T_Z, m	0.2345	0.2003
	X	−38.08	−30.9	$(1+\lambda)$	1.0002	0.9988
RS2	Y	8.31	19.30	ω_X	8' 16.2"	9' 22.1"
	Z	−14.39	15.62	ω_Y	5' 37.5"	4' 11.8"
	X	32.28	13.38	ω_Z	10' 28.4"	7' 43.1"
RS3	Y	29.04	7.28			
	Z	−14.39	15.62			
	X	42.23	−12.22			
RS4	Y	16.61	21.79			
	Z	4.11	−9.20			

From statistical test, the difference in monitoring points coordinates exceeds the expected surveying error at several points on the bridge and this is likely due to actual movement of these points. It is deduced that there are some monitoring points were moved (deformed) between the period from December 2011 to May 2012. The maximum value of deformation in this period was 8.3 cm in horizontal component and 4.7 cm in vertical direction. In the other hand, from the period December 2011 to June 2013, more points were moved.

8. CONCLUSIONS

Based on the results obtained, the following conclusions can be summarized:

1) The proposed geodetic monitoring techniques can provide valuable data on the deformation of the structural members and movement of structures such as highway bridges. The accuracy of the monitoring target coordinates will be improved if the two total stations are applied instead of using one total station only.

2) Application of seven parameters transformation technique is an effective method for determining the values of deformation of structure

because it can remove the systematic errors of observations between the two successive epochs;

3) Achieving the required accuracy for geodetic monitoring technique of highway bridge is based on the following factors:
    a. The used instruments specifications (Instrument resolution, data collection options and the proper operating instructions).
    b. The field observing and modeling procedures. Measurements and adjustment techniques of the network have direct influence on the detection of monitoring point's displacements;
4) All the recorded noise levels in all places around the studied highway bridge are much more the permissible limits according to the Egyptian law of environmental and world organizations limits;
5) Statistical test using F-Fisher criterion with a confidence level of 98% can be used to identify the values of deformation for any structure such as bridge between any two epochs of observations.

ACKNOWLEDGEMENTS

The author wants to thank students of Civil Engineering Department, Higher Institute of Engineering and Technology-Kafr El-Sheikh, Egypt for helping in the field observations collection.

REFERENCES

1. Beshr, A.A. (2004) Accurate Surveying Measurements for Smart Structural Members. M.Sc. Thesis, Public Works Department, Faculty of Engineering, Mansoura University, El-Mansoura.
2. Beshr, A.A. (2010) Development and Innovation of Technologies for Deformation Monitoring of Engineering Structures Using Highly Accurate Modern Surveying Techniques and Instruments. Ph.D. Thesis, Siberian State Academy of geodesy, Novosibirsk.
3. Vanatwerp, R.L. (1994) Engineering and Design-Deformation Monitoring and Control Surveying. US Army Corps of Engineering, Washington DC, USA.

4. Gairns, C. (2008) Development of Semi-Automated System for Structural Deformation Monitoring Using a Reflectorless Total Station. M.Sc. Thesis, Department of Geodesy and Geomatics Engineering, University of New Brunswick, Fredericton, Canada.

5. Lutes, J.A. (2002) Automated Dam Displacement Monitoring Using a Robotic Total Station. M. S. Thesis, Department of Geodesy and Geomatics Engineering, University of New Brunswick, Fredericton, Canada, 141.

6. Beshr, A.A. (2012) Monitoring the Structural Deformation of Tanks. LAP LAMBERT Academic Publishing, GmbH KG, Germany.

7. Roberts, G. and Dodson, A. (2001) A Remote Bridge Health Monitoring System Using Computational Simulation and GPS Sensor Data. Institute of Engineering Surveying and Space Geodesy.

8. Elsheimy, N. (2001) Adjustment of Observations. Department of Geomatics Engineering, University of Calgary.

9. (2002) Standard Specification for Highway Bridges. 17th Edition, American Association of State Highway and Transportation Officials, Washington DC.

10. Banerjee, D., Chakraborty, S.K., Bhattacharyya, S. and Gangopadhyay, A. (2008) Evaluation and Analysis of Road Traffic Noise in Asansol: An Industrial Town of Eastern India. International Journal of Environmental Research and Public Health, 5, 165-171.

11. (2002) Building Code Requirements for Structural Concrete. American Concrete Institute, Committee 318, Farmington Hills.

12. Yang, Y.M. and Myers, J.J. (2003) Live Load Test Results of Missouri's First High Performance Concrete Superstructure Bridge. Transportation Research Board, 82nd Annual Meeting, Washington DC.

13. Parbat, D.K. and Nagarnaik, P.B. (2007) Assessment and ANN Modeling of Noise Levels at Major Road Intersections in an Indian Intermediate City. Journal of Research in Science, Computing and Engineering, 4, 39-49.

11

CHAPTER

GEOTECHNICAL CHARACTERIZATION OF SAKAKINI'S PALACE STONES AND OTHER CONSTRUCTION MATERIALS, CAIRO-EGYPT

Sayed Hemeda

Conservation Department, Faculty of Archaeology, Cairo University, Cairo, Egypt

ABSTRACT

The understanding of the geotechnical problems and failure mechanisms of stone structures of Sakakini palace (1897 after century) entails a comprehensive study on the mechanical behaviour of the stones and other construction materials. In addition to micro analysis, geological and geomorphologic interests, several investigations on stone deterioration and engineering geology were performed, First phase included more sophisticated techniques, which provided additional information on particular aspects of site deterioration and it included laser induced breakdown spectroscopy (LIBs), electron probe micro analysis, micro XRD and XRF analyses, scanning electron microscope analysis coupled with EDX probing, transmission electron microscopy and grain size distribution analysis, permeability and pore size distribution of stone, mortars, core binders and other construction materials. Second phase included the determination of mechanical properties of building stones, such as compressive strength, modulus of elasticity, tensile strength, and shear strength. To arrive at reliable values for these properties, a suitable number of samples should be extracted, prepared for testing, and properly tested. The test results are then analyzed to establish the investigated stone properties. The testing program includes extracting seven cylindrical cores from the basement stone walls of Sakakini's mansion in down town Cairo. The cores are extracted using rotary cylindrical diamond blade coring machine. The top and bottom surfaces of every core were prepared to be flat circular surfaces perpendicular to the vertical axis of the core. Because the palace is museum and attractive places for the tourists, core sampling could be carried out only at a limited number of locations under official permission. For the purpose, cylindrical specimens with a diameter of 42 - 44 mm and height of 90 - 100 mm, prepared by the use of a core drilling machine and some collected blocks from the archaeological site under investtigation were taken to determine the bulk structure, physical, short and long-term mechanical properties of the stone and other construction materials in the laboratory. A number of specimens prepared from these blocks were employed for testing. Furthermore, limitation due to the number of blocks was overcome by the determination of the in situ characteristics of the stones by Schmidt hammer tests, geo-tomographic investigations and rock mass classification on some stone rock structures where testing has been permitted. The objectives of the study are to provide a characterization of micro structures and the mechanical properties of the stones of Sakakini's Mansion; describe the required testing plan; describe the test results and conclude the values of the basic mechanical properties of the building stones. The following sections provide detailed descriptions of the steps taken to achieve the objectives of the study. The purpose of the present research is to provide recommendations regarding the strengthening and the safety of architectural heritage under long and short-term loading. For this purpose, a set of experimental tests and of advanced numerical analyses are to be carried out.

Keywords: Architectural Heritage; Sakakini Palace; Geotechnical; Investigations; Mechanical Properties; Stone Deterioration; Laser Induced Breakdown Spectroscopy (LIBs)

1. INTRODUCTION

The site of Habib Sakakini's palace is considered from the important sites that have attracted the attention of a lot of historicans, especially at the Mamluk's and Ottoman's era. The location of that palace anciently was a water pond that was located at the eastern side of Egyptian gulf besiding Sultan Bebris Al-Bondoqdary mosque as that pond was so called "Prince Qraja al-Turkumany pond". That pond had acquired the extent of importance when it had been filled down by Habib Sakakini at 1892 After Century, where he had constructed his famous palace in 1897 AC, Figure 1.

Chalky stones and the other kind of construction materials (which are under investigation) are porous rocks with complex behaviour, two main deformation mechanisms can be commonly identified in theses rock types, depending on the applied stress conditions: 1) fracture propagation, associated with volumetric dilatancy and brittle behaviour, which is dominant in compressive stress paths at no or low confining pressure, 2) pore collapse, which dominates at higher stress conditions, producing large contracting plastic deformations.

The high fossils content, due mainly to shells of foraminifers and some mollusks, give rise to structural heterogeneity, which reflected in the variability of the mechanical properties and in the poor reproducibility of the experimental results [1,2].

There is no generally accepted theory of strength of brittle rock based on an examination of process of formation of micro cracks and deformation (Maizel.I.V and Sokolov.I.B), and establishing the initiation and development of stress-induced microfractures in the (EDZ) is thus of key interest.

Some of the key concerns regarding the stability of monumental structures of soft rocks, include the implications of potential ground disturbance by the constrction method and the redistribution of in situ stresses. Both of these factors relate to the initiation and propagation of brittle fractures and the extent of the excavation-disturbed zone (EDZ), which could adversely affect the stability of the stone structures boundary and could increase the permeability of the near-field host rock. In structural Geology and Tectonics, experimental rock deformation is important in determining the evolution of natural structures and tectonic features.

Figure 1. Sakakini palace in Cairo, Egypt.

Considerable effort has been extended towards the understanding of brittle fracture processes and mechanisms. Much of this focus has been extended to laboratory testing and the measurement/quantification of brittle fracture thresholds [3-7]. Of these the crack damage threshold (marked by the onset of dilatancy and taken as the point of reversal in the volumetric strain curve is of specific interest as several studies have associated the threshold with unstable crack propagation in brittle rocks [8]. unstable crack propagation corresponds to the point where the relationship between the applied stress and the crack growth velocity, take control of the propagation process. Under such conditions crack propagation will continue until failure even if the applied loading is stopped and held constant. As such, martin & chandler and read et al. [9] have equated the crack damage threshold to the long-term in situ strength of brittle rock, a practice they note that has been similarly adopted in the concrete industry.

Thus, the identification of these processes and their associated mechanisms are of key interest in predicting both the short and long-term strength of soft rock. This research focuses on these processes by presenting the results from several short and long-term laboratory tests.

In general, the propagation of a brittle fracture can be equated with the irreversible destruction of molecular cohesion along the generated fracture path. In this sense, the microfracturing process acts to "damage" the rock material. As the number of propagating fractures multiplies, damage can be viewed as accumulative and can be correlated to observed decreases in the elastic stiffness and cohesive strength of the material.

In this work, we highlighted some important characteristics of the geotechnical behaviour of structured soft rock and showed that these characteristics are quite common in many natural rocks. Based on such concepts, research into soft rock materials has intensified in the last two decades [10].

2. BULK STRUCTURE AND DURABILITY ASPECTS OF STONE, MORTARS AND CORE BINDERS

Various techniques were used to establish the causes of the stone deterioration in Sakakini palace. The first group of techniques was used to establish site parameters and it included photographic recording, petrographic analysis of the stone type, microclimate data, ground water composition and ultra violet radiation levels.

The second group included more sophisticated techniques, which provided additional information on particular aspects of site deterioration and it included laser analysis (LIBS) [11,12], electron probe micro analysis, micro XRD, Table 1 and XRF, Table 2 analyses, scanning electron microscope analysis coupled with EDX probing, transmission electron microscopy and grain size distribution analysis of rock samples, permeability and pore size distribution of stone, mortars and core binders.

Table 1. XRD analysis results for stone and other construction materials.

S. No.	Major Const	Minor Const	Trace Const
Chalky limestone	Calcite	-	Calcite
Mortars between stone blocks	Calcite	Calcite	Gypsum
Render or plaster layers	Quartz, Halite, Gypsum	-	Illite
Core binders between leave masonry walls	Quartz	Calcite	Calcite

Table 2. XRF analysis results for stone and other construction materials.

D.N	Chalky Limestone	Mortars between stone blocks	Render or Plaster layers	Core Binders between leave masonry walls
	%	%	%	%
SiO_2	0.81	63.28	30.18	65.28
TiO_2	0.05	0.25	0.06	0.55
Al_2O_3	1.13	3.9	0.63	7.27
Fe_2O_3	0.17	1.65	0.28	3.48
MnO	0.002	0.02	0.003	0.05
MgO	1.01	1.17	0.9	1.28
CaO	52.5	14.08	10.28	11.13
Na_2O	0.12	0.81	12.01	0.87
K_2O	0.15	0.73	1.94	0.88
P_2O_5	0.1	0.13	0.09	0.19
Cl	0.24	0.65	12.07	0.94
SO_3	0.9	1.85	11.34	1.76
L.O.I	42.45	10.83	20.1	5.97

2.1. Chalky Limestone

The limestone can be characterized as medium grained with uniform relative grain size, angular to sub angular grain shape with equidimensional form and rough surface texture. Sound or non weathered pieces of stone can be characterized of medium compactness and durability and the weathered pieces characterized of low compactness and durability. It must be mentioned that weathering attacked strongly the masonry units, started from the surface and continuing inward thus loosing the mineral fabric, Figure 2.

LIBS spectrum, Figure 2, and The EDX micro analyses indicated that, the elemental arrangement for the chalky limestone samples collected from Sakakini place can be put in a decreasing order according to their concentration as follow:

Ca (33.50%), O (29.80%), S (16.94%), K (9.01%), C (6.53%), Fe (1.76%), Si (1.48%), Al (0.55%), Mg (0.44%). Total 100.00% (Wt%). see Figure 3.

2.2. Mortars between Stone Blocks

Mortars between stone blocks are coarse to very fine grained, compact, and poorly sorted. It composed of quartz as the essential mineral constituents with rare amounts of rock fragments, hornblende, collophane, epidote, gluconite and opaques all cemented by very fine grained calcite (micrite) admixed with traces of iron oxides and clay materials. Quartz occurs as coarse to finegrains of rounded to subangular outlines, and some quartz grains are polycrystalline. Feldspars and rock fragments detected in the rock as coarse to medium grains.

LIBS spectrum and The EDX micro analyses indicated that, the elemental arrangement for the Mortars between stone blocks samples collected from Sakakini place can be put in a decreasing order according to their concentration as follow:

Ca (49.21%), O (32.70%), Si (6.77%), C (5.85%), Fe (2.25%), Al (2.25%), K (0.67%). Total 100.00% (Wt%).

2.3. Plaster or Render Layers

The renders layers are medium to fine-grained, compact. The renders composed mainly of quartz, Gypsum with rare amounts of feldspars, clay minerals and iron oxides. Quartz occurs as medium to fine-grains of rounded to subangular outlines, and some quartz grains are polycrystalline.

The matrix is composed of a mixture of micrite with minor amount of clay minerals and iron oxides.

The grade of disintegration and deterioration usually goes with the salt content, it was found that pollutants from car traffic, attack mortars as well as chlorines from the ground water, lower parts covered for long time with ground also present high percentage in Na Cl and NOand SO_4 salts.

LIBS spectrum and The EDX micro analyses indicated that, the elemental arrangement for the renders layers samples collected from Sakakini place can be put in a decreasing order according to their concentration as follow:

O (37.50%), Si (26.60%), C (11.72%), Ca (10.42%), S (9.52%), Mg (1.23%), Na (1.21%), Cl (1.1%), Al (0.71%). Total 100.00% (Wt%).

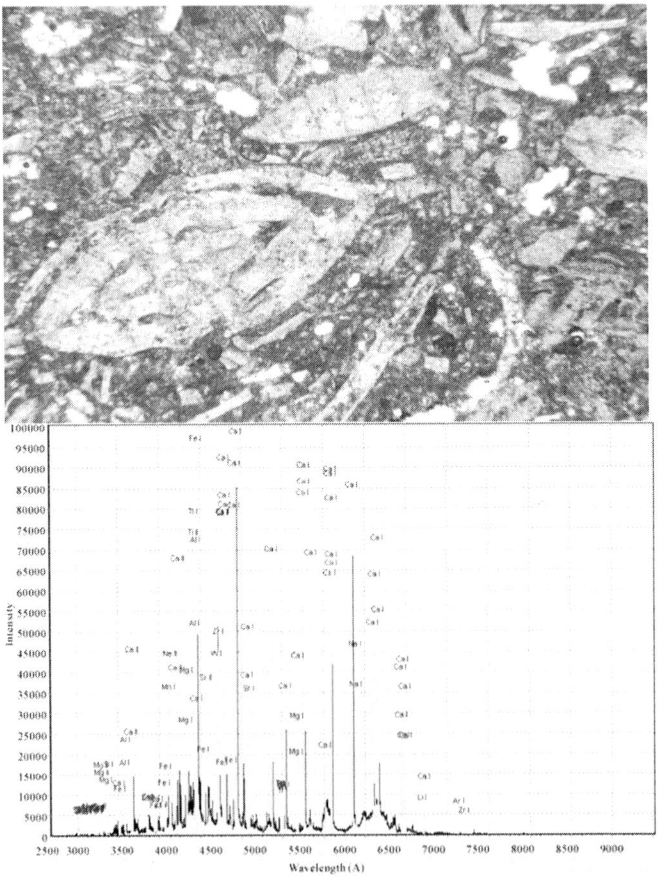

Figure 2. Thin section and Libs spectrum for an analysis spot on building limestone.

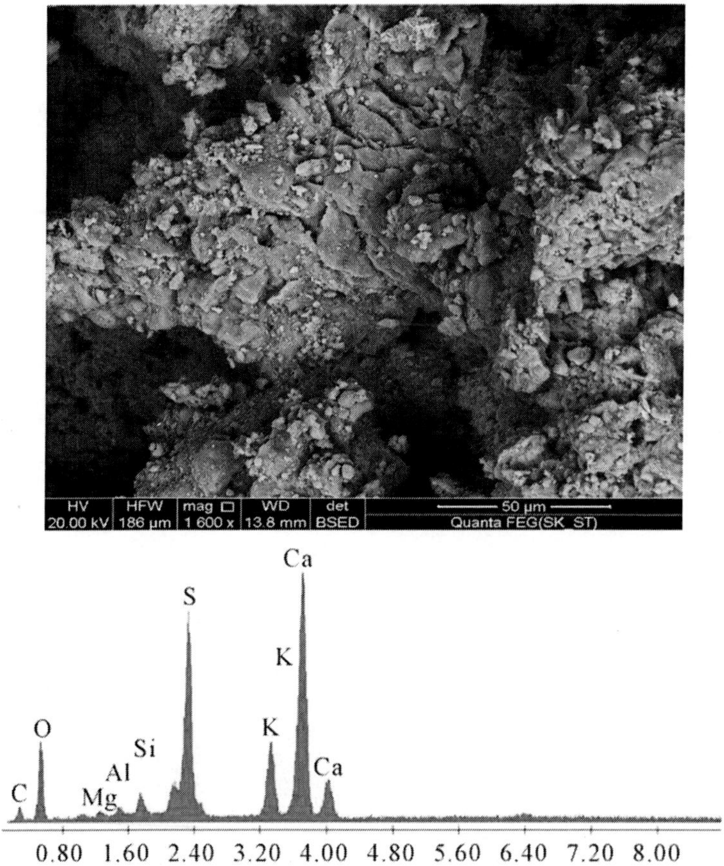

Figure 3. SEM with EDX analysis results of chalky limestone.

2.4. Core Binders between Leave Masonry Walls

The core binders between two leaves are coarse to veryfine grained compact. It composed of quartz as the essential mineral constituents-with minor amount of feldspar and rare amounts of rock, stone and brick fragments, collophane, glauconite and opaques all cemented by very fine grained calcite (micrite) admixed with traces of iron oxides and clay materials. Quartz occurs as coarse to fine-grains of rounded to subangular outlines, and some quartz grains are polycrystalline. Feldspars and rock fragments detected in the rock as coarse to medium grained The grade of disintegration and deterioration usually goes with the salt content. Although the research was not oriented to estimate the grade of deterioration, it was found that pollutants from car traffic, attack mortars as well as chlorines from the ground water, lower parts covered for long time with ground also present high percentage in NaCl and Na_2SO_4 salts.

LIBS spectrum and The EDX micro analyses indicated that, the elemental arrangement for the Core binders between leave masonry walls samples collected from Sakakini place can be put in a decreasing order according to their concentration as follow:

Ca (39.29%), O (29.44%), Si (17.13%), C (5.17%), Al (4.73%), Fe (3.16%), S (9.52%), Cl (0.58%), Mg (0.49%). Total 100.00% (Wt%).

3. MECHANICAL TESTING PROGRAM

The basic mechanical properties of building stones are compressive strength, modulus of elasticity, tensile strength, and shear strength. To arrive at reliable values for these properties, a suitable number of samples should be extracted, prepared for testing, and properly tested. The test results are then analyzed to establish the investtigated stone properties.

The testing program includes extracting seven cylindrical cores from the basement stone walls of Sakakini's mansion in down town Cairo. Figure 4 The cores are extracted using rotary cylindrical diamond blade coring machine. The top and bottom surfaces of every core were prepared to be flat circular surfaces perpendicular to the vertical axis of the core, Figure 5.

Three cores were equipped with strain gauges to measure the longitudinal strain corresponding to compressive stresses. The results of testing these three cores in compression will be used to determine the stones compressive strength and modulus of elasticity.

Three cores were tested in splitting tension to determine the tensile strength of the stone samples. One core was tested in single shear to determine the stone's shear strength.

The following sections describe the different tests and discuss their results.

Figure 4. Core samples extracting from the basement.

Figure 5. Some core samples after preparation.

3.1. Compressive Strength

Three samples were equipped with 100 mm long electric strain gauges. One strain gauge was fixed to the core parallel to its vertical axis to measure the longitudinal compressive strain corresponding to the applied compressive stress, Figure 6. The vertical compressive stress was increased gradually until failure. The compressive strength of every core was determined. Table 3 shows the results of the compressive strength of the three cores, Figures 7 and 8.

3.2. Modulus of Elasticity

The recorded stress-strain data for the three samples are graphed as shown in Figures 9-11. From these figures, the value of the modulus of Elasticity can be obtained. For brittle materials with non-linear stress-strain relationship; the modulus of elasticity may be determined as the chord modulus between a low stress value (0.5 MPa) and a stress value equals one third of the compressive strength. For such definition the values of the modulus of elasticity obtained for the three tested samples are given in Table 4.

3.3. Splitting Tensile Strength

Three samples were tested to determine the average splitting tensile strength of the stone. Figure 12 shows a sample in the test set-up at failure. Table 5 gives the details of the test results.

3.4. Shear Strength

One sample was tested in single shear to determine the shear strength of the stone. The obtained shear strength equals 1.63 MPa.

3.5. Abrasion Test

Results from the abrasion test on some white Karara Marble specimens indicated that, the loss of weight is about 7-8 g, and loss of thickness is 0.66 mm, Table 6.

Figure 6. Test set-up.

Figure 7. Compressive strength sample at failure.

Figure 8. Compressive strength sample in the test set-up.

3.6. Bending Moment

Bending moment test on some Boghdadi blocks wooden specimens had been carried out and the results indicated that the bending moment value ranging between 379 and 592 kg/cm^3, Table 7.

Table 3. Compressive strength test results.

No.	Diam. (mm)	Height (mm)	Cross Sectional Area (mm^2)	Failure Load (kN)	Compressive Strength (MPa)	Remarks
1	44.7	132	1569.3	16.50	10.5	-
2	42.7	117	1432.0	14.90	10.4	-
3	44.1	136	1527.5	16.22	10.6	-
4	49.1	140	1886.3	17.80	9.44	-
5	49.1	161	1886.2	22.90	12.14	-
6	49.1	168	1886.2	13.30	7.05	-
7	49.1	200	1886.2	15.70	8.32	-

Table 4. Modulus of elasticity for the stone samples.

No.	Strain at 0.5 MPa ($\times 10^{-6}$)	Ultimate Strength f_{ult} (MPa)	Strain at $f_{ult}/3$ ($\times 10^{-6}$)	Modulus of Elasticity (GPa)	Remarks
1	10	10.5	301	17.182	-
2	13	10.4	213	22.500	-
3	4	10.6	250	18.293	-

Table 5. Splitting tensile test results.

No.	Diam. (mm)	Height (mm)	Ultimate Load (kN)	Splitting Tensile Strength (MPa)	Remarks
1	44	101	10.35	1.48	-
2	44	117	15.02	1.86	-
3	44	117	12.07	1.49	-
4	49	12.2	10.90	1.16	-
5	49	11.0	15.00	1.77	-
6	49	13.6	11.70	1.12	-
7	49	13.2	13.57	1.34	-

3.7. Physical Properties

Generally speaking, it could be said that although high absorption means low strength there are some exceptions. Where we have high absorption we have high strength and of course, for the same strength a high range of absorption. Helped by the record card data bank files, these peculiarities can be explained. High content in clay or argillaceous materials increase absorption. This obvious in microscopic analysis.

When cracks are present in the mass of the building stones the absorption is over 0.6% Table 8. For the sound stone samples absorption ranges from 0.4% to 0.5%. For 50% of the samples absorption is below 0.5%.

Table 6. Abrasion Test Results of Marble specimens.

No.	Dry Weight (gr)	Loss of Weight (gr)	Loss of Thickness (mm)	Remarks
1	211	7	0.66	White Karara
2	213	7	0.56	-
3	215	8	0.70	-

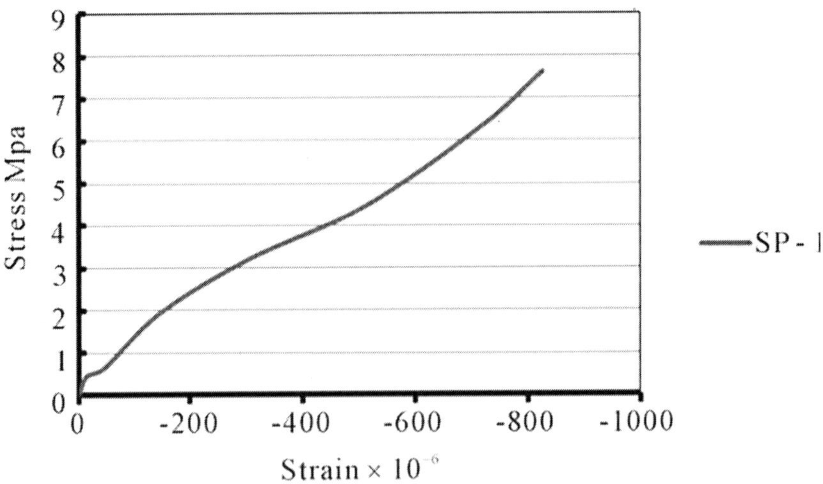

Figure 9. Stress-strain curve for sample 1.

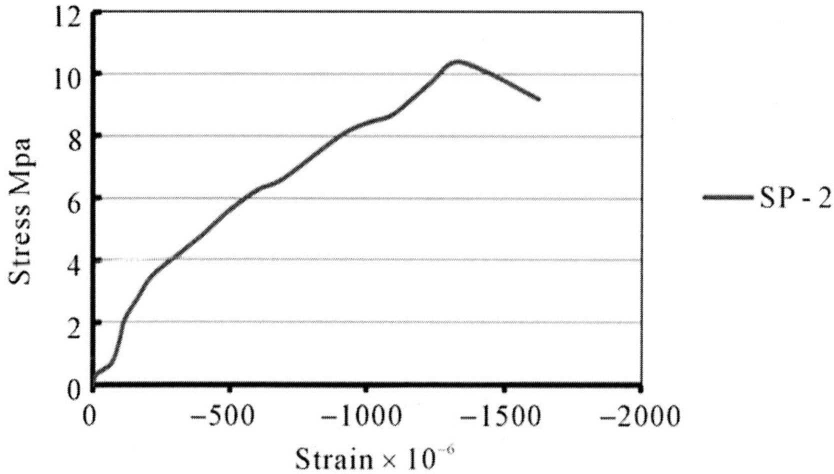

Figure 10. Stress-strain curve for sample 2.

For the 65% of the samples the apparent specific gravity ranges from 1.8 to 2 gr/cm³.

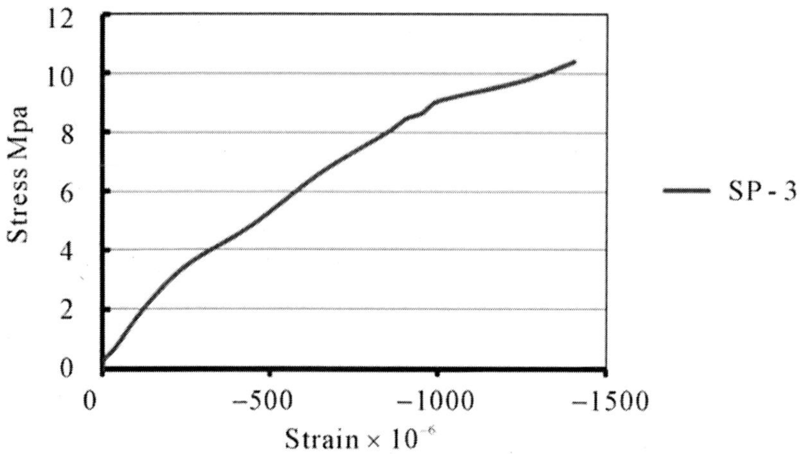

Figure 11. Stress-strain curve for sample 3.

Figure 12. Splitting Tensile Test Sample at Failure.

Table 7. Bending moment test results of wooden specimens.

No.	Dimensions (mm)	Span (mm)	Ultimate load (kg)	Bending moment (kg/cm^2)	Remarks
1	94 × 31	20	1140	379	Block
2	30 × 13	20	100	591.7	Boghdadi
3	30 × 13	20	100	592	-

Table 8. Physical properties of marble specimens.

No.	Dry weight (gr)	Saturated weight (gr)	absorption (%)	Remarks
1	124	124.5	0.4	White Karara
2	124	125	0.4	-
3	124	125	0.5	-

4. PATHOLOGY ASSESSMENT AND CAUSES OF DETERIORATION

The structure of the vast complex of the palace is unusually complicated. These buildings are of non-homogenous structural features, their elements are made of different construction materials and its foundations are of various levels and configurations.

An inspection of the palace revealed that while the deterioration seems to be geotechnical, seismic and moisture/weathering problems, other instances of deteriorations are caused by structural systems of the buildings. The structural system is a largely to be regarded as a free shear-wall system, in which the load-bearing walls are neither always placed on top of each other nor upon strong beams transmitting their load to other continuous structural components.

The foundations were found to be built on different levels. Under a certain wall, the footing was found to be partly supported on the remains of the underground structure while the other parts are in the fill layer. The foundations in some locations were not found. Visual cracks were observed to verify that the foundations of the building had been subjected to differential settlements.

The stability problems of these structures are mainly related to the damage caused by the earthquake & seismic and geotechnical hazards also the foundation subsidence and deformations and fractures on the walls, as well as the weathering of stones and others construction materials by ground water, polluted atmosphere, temperature variations and the mechanical action of wind, where historical building stones exposed to atmospheric factors are subject to the action of various agents causing their destruction. A complex combination of mechanical, chemical, biological factors played an important role in the deterioration of the building stones of palace.

Type of damage recognized:

1) Flakes or scaling at the surface 2) Loss of mass 3) Changes in shape (worn rectangular angles)

4) Efflorescence 5) Changing in colour (smoked brick)

6) Presence of fungus 7) Pulverization The symptoms of building stones disintegration usually go in parallel with the high humidity of the walls. Salts are often concentrated in the voids without cracking around them. In many cases recrystallization takes place in the cracks which constitute the veneers of liquid faces. Salt crystals are usually in microcrystalic form and rarely in well developed crystal shape. It seems that the high content in salts has a detrimental effect on stone strength.

The process of the mechanical deterioration affecting the structures are abrasion, polishing, scouring, scratching splitting, cracking fragmentation and stone fall. These processes are reflected by striations, polished surfaces, undulating surfaces, cracks, fissures and joints and finally rock loose. Water/salt combination activated by fluid circulation and evapo-transpiration, is one of the main factors which provoke weathering and decomposition of a stone, by reaction of shrinking-expansion phenomena of pore water.

We can see decays by supply of matter such as brown crust, damages by loss of matter such as dissolution, other degradations by loss of matter such as granular disintegration and contour scaling (on parts exposed to rains), the presence of salts leads to dissolution and crystallization cycles of salts, especially inside the porous network, this mechanism occurs: 1) by capillary transfer of water and salt from the ground and 2) by fixation of water vapor by sea salt deposited on the stone by the wind. And the action of plant's root in a particularly porosity material of construction.

The role of human activities is very dangerous in which the building stone were subjected to scratch, breaking down or removing and reuse of stones.

5. CONCLUSION

A testing program was planned and executed to determine the basic structure/composition and mechanical properties of Sakakini's Mansion

Building Stones. Micro structure analysis, the compressive strength, Modulus of Elasticity, Splitting tensile Strength and Shear Strength were determined as shown in the previous sections of this report.

6. ACKNOWLEDGMENT

The present paper is part of project entitled "risk assessment and seismic response analysis of architectural heritage in Egypt" funded by Cairo University and under the management of Dr Sayed Hemeda, the author of the paper and main investigator of the project. I gratefully acknowledge the support and assistance of Professor Osama A. Hodhod, professor of strength of materials, Structural Engineering Department, Faculty of Engineering, Cairo University for his great efforts and help through the investigations, while the mechanical testing had been carried out in the laboratory of (CERC) Civil Engineering Research Centre, Cairo University under request and supervision of Dr Sayed Hemeda.

REFERENCES

1. Z. S. Bieniawski, "The Geomechanics Classification in Rock Engineering Applications," 4th ISRM Congress, International Society for Rock Mechanics, Switzerland, 1979, pp. 41-48.
2. M. Bukovansky, D. P. Richard and K. R. Week, "Influence of Slope Deformations on the Tombs in the Valley of the Kings, Egypt," Proceedings of an International Symposium on Engineering Geology and the Environment, Vol. 3, 1997, pp. 3077-3080.
3. D. U. Deere and O. A. Varde, "General Report, Engineering Geological Problems Related to Foundations and Excavations in Weak Rocks," Proceedings of the 5th International Association of Engineering Geology Congress, Vol. 4,1990, pp. 2503-2518.
4. A. De Meer and C. J. Spiers, "Creep of Wet Gypsum Aggregates under Hydrostatic Loading Conditions," Tectonophysics, Vol. 245, No. 3-4, 1995, pp. 171-183.
5. A. De Meer and C. J. Spiers, "Uniaxial Compaction of Wet Gypsum Aggregates," Journal of Geophysical Research, Vol. 102, No. B1, 1997, pp. 875-891.

6. M. Cecconi and G. Viggiani, "Physical and Structural Properties of a Pyroclastic Soft Rock," In: Evangelista and Picarelli, Eds., Proceedings of the Geotechnics of Hard Soils-Soft Rocks, Rotterdam, 1998, pp. 2320-2331.
7. M. R. Cooper, "The Mechanics of Uncemented Carbonate Sands," Geotechnique, Vol. 40, No. 4, 1990, pp. 607-627.
8. L. S. Costin, "Time-Dependent Deformation and Failure," In: B. K. Atkinson, Ed., Fracture Mechanics of Rock, San Diego, 1987, pp. 167-215.
9. S. Hemeda, K. Pitilakis, S. Bandis, I. Papayianni and M. Gamal, "The Underground Monuments (Catacombs) in Alexandria, Egypt," Proceedings of the 4th International Conference on Earthquake Geotechnical Engineering, Thessaloniki, 25-27 June 2007, pp. 715-738.
10. S. Hemeda, "An Integrated Approach for the Pathology Assessment and Protection of Underground Monuments in Seismic Regions. Application on some Greek-Roman Monuments in Alexandria, Egypt," Ph.D. Thesis, Aristotle University of Thessaloniki, Thessaloniki, 2008.
11. C. Cennini, "Il libro dell'arte," Neri Pozza, ISBN: 88730- 591041, 2003.
12. D. L. Massart, B. G. M. Vandeginste, S. N. Deming, Y. Michotte and L. Kaufman, "Chemometrics: A Textbook," Elsevier Science Publishers, Amsterdam, 1988.

12
CHAPTER
SEISMIC EVALUATION OF STRUCTURAL INSULATED PANELS IN COMPARISON WITH WOOD-FRAME PANELS

Stefanie Terentiuk [1] and Ali Memari [2,*]

[1] *Structural Engineer, Zieman Engineering, LLC, Stamford 06901, CT, USA*

[2] *Department of Architectural Engineering, The Pennsylvania State University, University Park 16802, PA, USA*

ABSTRACT

Structural Insulated Panel (SIP) wall systems have been used in residential and light commercial buildings for the past sixty years. Lack of sufficient published research on racking load performance and limited understanding of the influence of fastener types on seismic response has been a deterrent in widespread use of the wall system in seismically active areas. This paper presents the results of a study involving a total of twenty one 2.4 m × 2.4 m shear walls tested under monotonic and cyclic loading. Four different 114 mm thick SIP panel configurations and one traditional wood frame wall were tested under monotonic loading according to ASTM E 564-06; and thirteen 114 mm thick SIP panels and three wood frame walls were tested under the CUREE loading protocol according to ASTM E 2126-11.

Parameters such as fastener type; spline design; hold-down anchor location; and sheathing bearing were adjusted throughout the testing in order to determine their effects on the SIP's performance. Performance parameters such as peak load and displacement; energy dissipation; allowable drift load capacity and seismic compatibility were determined for all of the specimens. Such parameters were then used to demonstrate the SIP walls' compatibility with the wood frame walls and to determine the efficiency of the different SIP wall configuration and spline systems employed.

Keywords: structural insulated panels (SIP); seismic evaluation; racking test; wood panels; residential construction

1. INTRODUCTION

In recent years, green building construction has significantly increased the demand for Structural Insulated Panels (SIPs). SIPs are usually constructed using two sheets of plywood or oriented strand board (OSB) with a rigid foam insulation core of expanded polystyrene (EPS) or extruded polystyrene (XPS), polyisocyanurate, or polyurethane. Typically, SIPs are used as load bearing prefabricated wall systems. Individual SIP panels are joined together in the field with spline and connection hardware determined by the manufacturer. Although extensive research has been performed on lateral load resistance of wood-frame wall systems, there have been relatively few published experimental parametric studies on SIPs. Furthermore, any tests that have been performed are often undertaken by SIP manufacturers, so the results are typically considered proprietary information and therefore not available to the general public. Limited availability of extensive technical data and background information may deter contractors, engineers, and homeowners from using SIP products, and even more so in seismically active regions.

The primary objective of the pilot study [1] presented in this paper was to gain a better understanding of how SIPs perform under in-plane lateral loads such as those induced by seismic events or experienced during high wind loading. Parameters such as spline design, connection hardware, hold-down methods and sheathing bearing were evaluated with the use of a full-scale testing program in order to determine their effect on the SIP wall system and subsequent performance under cyclic racking. Two common SIP spline designs were tested while the connection hardware was kept constant. Then, the three most commonly used connection hardware were tested, while the spline design was held constant. This paper presents a brief literature review, explanation of the testing program, description of the specimens, discussion of the test results, and presentation of seismic evaluation of the tested specimens based on ICC-ES guidelines [2,3,4] and NTA procedures [5,6].

2. LITERATURE REVIEW

One of the few published full-scale test studies on the performance of SIPs under in-plane shear loading is presented by Jamison [7] who performed monotonic and cyclic tests on several 2.4 m × 2.4 m SIP specimens. Mosalam and Günay [8] tested full-scale SIP specimens under cyclic loading while adjusting parameters such as nail spacing, gravity loading, loading protocol (CUREE vs. Hybrid Simulation), ground motion type and the presence/lack of an analytical substructure. In the APA Report T2006P-33 [9] Premier Building Systems' SIPs were tested under racking shear, axial loading, and transverse loading. Kermani and Hairstans [10] tested SIPs under racking loads and combined bending and axial compression in order to determine the effects size and location of an opening have on the performance of a SIP assembly. Some studies have found that sealants and adhesives have an effect on the performance of SIP and wood frame shear walls [11]. Manufacturers typically have their own type of sealants and adhesives as well as their own application methods. In order to make sure the results of this research are useful to the industry and compliant with ICC-ES AC04 [3], the sealants and adhesives were left out. The results will be conservative regardless of the type of manufacturing method used.

Most SIP manufacturers have ICC-ES Reports that document the allowable loads for their products. The ICC-ES Legacy Report for Insulspan [12] reports the allowable racking load for panels with stapled plywood surface splines and for nailed wood splines. Insulspan's Technical Bulletin No. 111 [13], provides the shear strength of SIPs with varying spline designs and nail spacing. The ICC-ES Legacy Report for Intermountain Building Panels L.L.C. [14] tested panels with metal stud splines connected to the panels with screws. In the R-Control Tech Bulletin [15], 2.4 m × 2.4 m SIP walls with hold-downs were tested under the Structural Engineering Association of Southern California (SEAOSC) loading protocol [16]. Architectural Testing, Inc. [17] followed ICC-ES AC04 [2] to test Agriboard Industries' 2.4 m × 2.4 m compressed agricultural fiber sandwich panels under the SEAOSC loading protocol. Johnston et al. [18] and Lebeda et al. [19] both studied the effects hold-down anchors have on wood shear walls under the CUREE loading protocol.

A common trend amongst the majority of SIP manufacturers listed above is their lack of publicly available information concerning their panel design's performance under cyclic loading. The monotonic shear strength can be used to determine a material's reaction to wind loading but it does not directly correlate to the system's performance under seismic loading. R-Control [15] and Agriboard Industries [17] have tested their panels under cyclic loading but their reports lack a parametric analysis of the design methods comparing spline design, connection hardware, or hold-down methods. Johnston et al. [18] and Lebeda et al. [19] both studied the effects hold-down anchors have on wood shear walls under the CUREE loading protocol.

3. DESCRIPTION OF SPECIMENS

Six different SIP specimen configurations and one traditional light-frame wood wall were tested in this study (Table 1). The SIPs were 114 mm thick with an 89 mm expanded polystyrene core and two 11 mm OSB facings. Figure 1 shows a perspective view of a typical SIP panel with exposed components of the OSB skin, top and bottom plates, rigid core insulation, and interconnecting splines. The two spline designs examined in this paper include an OSB surface spline and a double 38 mm × 89 mm lumber spline as shown in Figure 2. In addition to testing the vertical joint configurations, three different types of connection hardware were tested. To mimic current practices, the fasteners included 8 d common nails, No. 6 plywood/particleboard screws, and 16 gauge staples. The fasteners were spaced 152 mm on-center (o.c.).

The base SIP test setup included two 1.2 m × 2.4 m panels joined with a surface spline along the 2.4 m vertical side of the panel. Fasteners spaced at 152 mm o.c. and Spruce Pine Fir of Grade 2 or better were used to frame the 2.4 m × 2.4 m specimen. The top and bottom plates were 38 mm × 89 mm placed within the OSB sheathing of the SIP. An additional 38 mm × 89 mm was placed on the top and bottom of the wall as shown in Figure 3 to prevent bearing between the sheathing and the test setup. Common practice in the field is for the sheathing to bear directly on the sill plate, but it is conservative to test the walls with non-bearing sheathing. The double end posts were connected to each other and to the top and base plates with 16d common nails spaced at 610 mm o.c. (per 2006 International Building Code [20]) and 102 mm from each end. USP PHD6 hold-downs were

Table 1. Test matrix of specimens tested.

Panel Type	Panel to Panel Connection	Bottom Plate	Top Plate	End Posts	Fastener Hardware	Fastener Spacing o.c.	Bearing	External Hold-down
A1 Includes: A1-1M, A1-1C, A1-2C	11.1 mm × 76 mm OSB surface spline	(1) 38 mm × 89 mm	(1) 38 mm × 89 mm	(2) 38mm × 89 mm	8 d common nail (3.3 mm Φ) 63.5 mm long	152 mm	No	Yes
A1 Bearing-3C	11.1 mm × 76 mm OSB surface spline	(1) 38 mm × 89 mm, (1) 38 mm × 140 mm sill plate	(1) 38 mm × 89 mm, (1) 38 mm × 140 mm top plate	(2) 38mm × 89 mm	8 d common nail (3.3 mm Φ) 63.5 mm long	152 mm	Yes	Yes
A1 Internal-4C	11.1 × 76 mm OSB surface spline	(1) 38 mm × 89 mm	(1) 38 mm × 89 mm	(2) 38 mm × 89 mm	8 d common nail (3.3 mm Φ) 63.5 mm long	152 mm	No	No (internal hold-down)
A3 Includes: A3-1M, A3-1C, A3-2C	11.1 mm × 76 mm OSB surface spline	(1) 38 mm× 89 mm	(1) 38 mm × 89 mm	(2) 38 mm × 89 mm	16 ga. staple (1.6 mm Φ) 38.1mm long	152 mm	No	Yes
A4 Includes: A4-1M, A4-1C, A4-2C, A4-3C	11.1 mm × 76 mm OSB surface spline	(1) 38 mm × 89 mm	(1) 38 mm × 89 mm	(2) 38 mm × 89 mm	No. 6 screw (3.5 mm Φ) 31.8 mm long	152 mm	No	Yes
B Includes: B-1M, B-1C, B-2C, B-3C	Double 38 mm × 89 mm spline	(1) 38 mm × 89 mm	(1) 38 mm × 89 mm	(2) 38 mm × 89 mm	8 d common nail (3.3 mm Φ) 63.5 mm long	152 mm	No	Yes
C Includes: C-1M, C-1C, C-2C, C-3C	Built as a 2.4 m × 2.4 m timber wall without a splice	(1) 38 mm × 89 mm	(2) 38 mm × 89 mm	(2) 38 mm × 89 mm	8 d common nail (3.3 mm Φ) 63.5 mm long	152 mm ext. 304 mm int.	No	Yes

Note: The No. 6 screws are C1018-C1022 steel with a minimum Rockwell Hardness of C44.

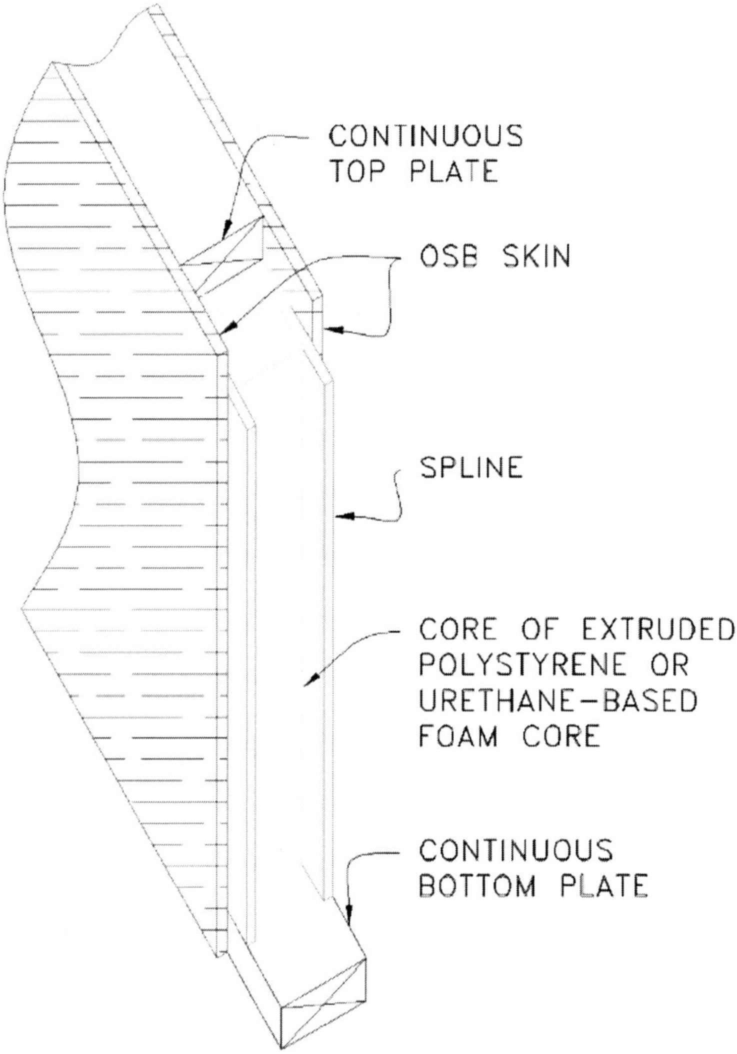

Figure 1. Typical SIP panel configuration. Adapted from [21].

attached to the outside of the SIP as shown in Figure 3. Refer to Table 1 for an explanation of the design specifications followed for each wall design. Further explanation of the testing setup and specimens can be found in Terentiuk [1].

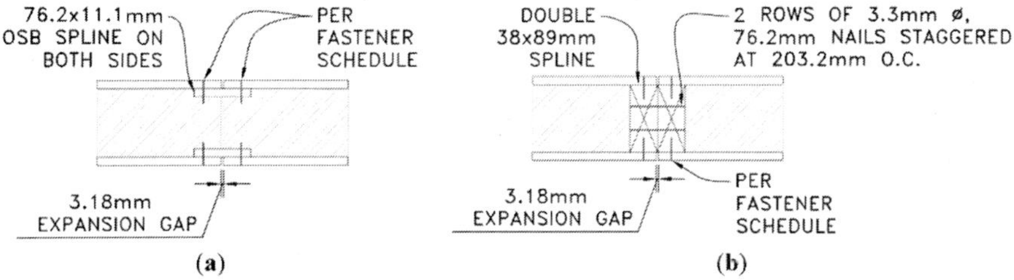

Figure 2. (a) OSB surface spline; (b) Double 38 mm × 89 mm lumber spline.

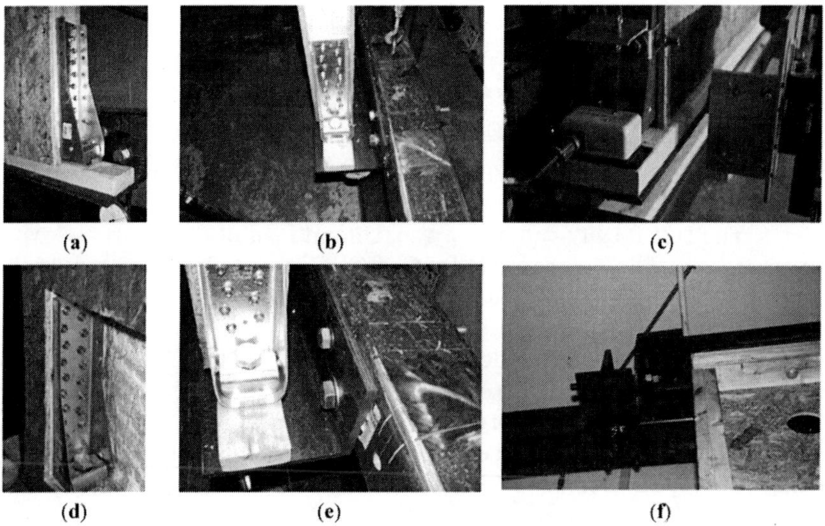

Figure 3. (a) Extra 38 mm × 89 mm base plate used to prevent friction between sheathing and test facility; (b) USP PHD 6 hold-down anchor located on exterior of panel; (c) 38 mm × 140 mm sill plate used in Specimen A1 Bearing to create sheathing bearing along top and bottom of panels; (d) Interior USP PHD 6 hold-down fit into 342.9 mm × 393.7 mm cut-out in SIP panel of Specimen A1 Internal; (e) 25.4 mm Φ screws used to attach L203 mm × 152 mm × 13 mm to bottom sliding steel tube; (f) MC8×20 attached to top of specimen and sliding connection used for application of load.

The traditional light-frame wood wall, Specimen C, was sheathed on both sides with 11 mm OSB oriented vertically. The OSB was taken from the same batch used to make the SIP specimens tested in this study. 38 mm ×

89 mm Spruce Pine Fir of Grade 2 or better was used for the studs and placed at 406 mm o.c. The specimen had a double 38 mm × 89 mm top plate, single 38 mm × 89 mm base plate and double 38 mm × 89 mm end posts. The sheathing was attached to the framing with 8d common nails spaced 152 mm o.c. edge and 305 mm o.c. field. The nailing patterns outlined in the 2006 IBC [20] were followed to connect the framing members with 16d common nails. Consistent with the SIP specimens, bearing of the sheathing was prevented with an additional 38 mm × 89 mm on the top and base of the wall and the hold-down anchors were placed on the exterior of the specimen.

4. TEST SETUP AND PROCEDURE

The Dynamic Racking Facility in the Building Components and Envelopes Research Lab at The Pennsylvania State University was used to test the wall systems. The test facility can apply a maximum load of 88,960 N and can displace a total of 152 mm. ASTM E 564-06 [22] was followed to test each specimen under monotonic loading and used to determine the load-displacement relationship of the wall systems (as noted in ICC-ES AC130 [4]). The loading protocol developed as part of the CUREE-Caltech Woodframe project [23] was used in this study to subject the specimens to cyclic loading, similarly to previously published testing and research [24,25]. The testing was carried out in a deformation controlled mode. More details on the cyclic loading for the tests are presented by Terentiuk [1] and Terentiuk and Memari [26].

ASTM E 2126 [27,28], ICC-ES AC130 [4], and ICC-ES AC04 Appendix A [2,3] were used to evaluate the SIP wall systems and compare the effects the parameters have on a SIP's behavior. The shear resistance of the SIP wall system was also compared to that of a conventional timber frame under static monotonic and cyclic loading.

5. EXPERIMENTAL RESULTS

5.1. Specimen Test Results

Table 2 describes the failure mode of each of the panels tested. The order in which the specimens are listed in the table and in the following paragraphs is based on increasing load resistance capacity, determined by this study. The initial failure of Specimen A3 occurred along the spline when the staples pulled out and sheared. At that point, the load was transferred to the sheathing connection along the top plate and base plate. Figure 4a shows the separation of the top plate from the sheathing as the staples sheared along the top and the nails pulled out of the end posts.

Table 2. Specimen failure modes.

Specimen	Failure Mode
A3 Monotonic	Staple withdrawal along base plate and top plate, and tear-out damage to sheathing along spline.
A3 Cyclic	Staple shear along spline and subsequent staple shear and withdrawal along base and top plate.
A4 Monotonic	Screw shear along spline and subsequent screw shear along top and base plate.
A4 Cyclic	Screw shear along spline and subsequent screw shear along top and base plate. Top plate pulled away from sheathing and end posts.
A1 Monotonic	Deformation and initial withdrawal of nails. Sheathing damage on inner corners of panels (along spline).
A1 Cyclic	Initial nail withdrawal along spline and top and bottom plates. Sheathing damage on inner corners of panels.
A1 Bearing Cyclic	Initial nail withdrawal along spline and top and bottom plates. Sheathing failure on inner corners of panels along top plate.
A1 Internal Cyclic	Initial nail withdrawal along spline. Sheathing damage along inner corners of panels and along base plate below hold-down cut-out.
B Monotonic	Deformation and initial withdrawal of nails along top plate. Sheathing damage on inner corners of panels. Double 38 mm × 89 mm pieces in lumber spline began to split apart from each other.
B Cyclic	Double 38 mm × 89 mm pieces in lumber spline began to split apart from each other. Nail withdrawal along base plate and tear-out sheathing failure along top plate.
C Monotonic	Initial nail withdrawal along spline and slight sheathing damage on inner corners of panels.
C Cyclic	Initial nail withdrawal along spline, base plate and end posts. Sheathing damage on inner corners of panels along top plate.

Similar to Specimen A3, the failure mode of Specimen A4 occurred in the fastener hardware along the spline. The brittle nature of the screws caused them to shear which resulted in a sudden and brittle failure. Figure 4b,c show the separation of the SIP panels caused by fastener hardware and subsequent base plate and top plate damage. Nail withdrawal along the spline caused the initial failure in Specimen A1. Rotation of the panels caused damage to the sheathing along the inner corners of the panel, refer to Figure 4d,e. Due to bearing of the sheathing along the sill plate, Specimen A1 Bearing experienced more extensive damage to the sheathing along the top and base plates than Specimen A1, as seen in Figure 4f. Placing the hold-down anchor on the interior of the panel in Specimen A1 Internal did not result in a significant difference in the failure mode of the specimen in comparison to Specimen A1. Figure 4g,h show sheathing damage at the inner corners of the panels and nail pullout along the spline and top plate of Specimen A1Internal. Under cyclic loading, the major failure in Specimen B occurred when the double 38 mm × 89 mm spline split apart vertically along the nailed connection. The 38 mm × 89 mm pieces pulled away from each other and remained relatively connected to the individual SIP sections. As a result, the two panels began to rotate independently of each other as seen in Figure 4i. The initial signs of failure for Specimen C occurred when the nails withdrew along the spline, base plate and end posts. Figure 4j,k show the nail withdrawal along the end post of Specimen C after cyclic loading. Figure 4k also shows the slight damage which occurred to the end posts.

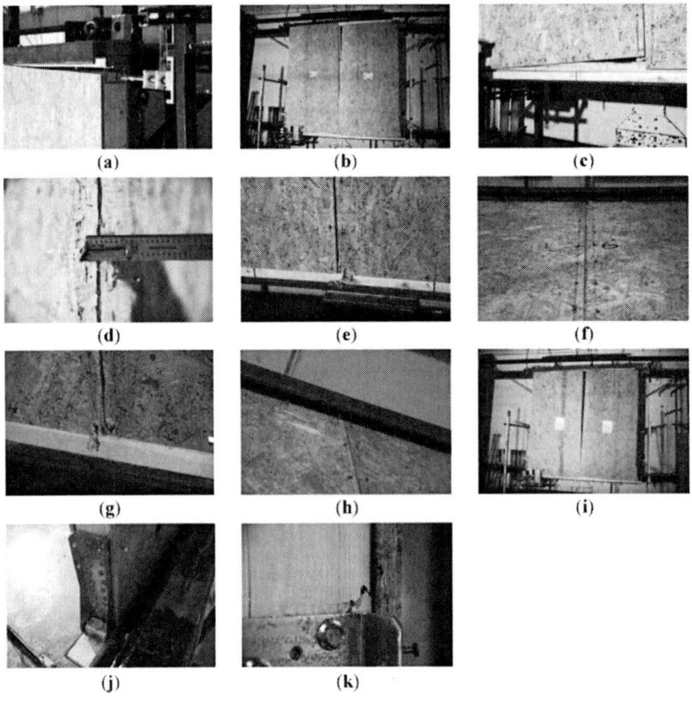

(a) (b) (c) (d) (e) (f) (g) (h) (i) (j) (k)

Figure 4. (a) Specimen A3 staple shear and top plate displacement; (b) Specimen A4 separation of panels along vertical spline; (c) Specimen A4 damage at base plate and screw shear along spline and base plate; (d) Nail withdrawal along spline of Specimen A1; (e) Sheathing damage along base plate of Specimen A1; (f) Pull-out of nails along spline and damage to sheathing along top plate of Specimen A1 Bearing; (g) Sheathing damage at inner corner of panels along base plate of Specimen A1 Internal; (h) Pull-out of nails along spline and top plate of Specimen A1 Internal; (i) Specimen B separation of panels (more extreme than the failure of two other identical walls); (j) Specimen C nail withdrawal along end post; (k) Damage to the end post of Specimen C.

5.2. Fatigue Testing

In this paper, the word "fatigue" refers to repeatedly loading a specimen under the first 37 cycles of the CUREE cyclic loading protocol. After a specimen was loaded under the first set of loading cycles, it was brought back to a displacement of zero. Without repairing the specimen in any way, the wall was run through the first 37 cycles of the CUREE cyclic loading protocol a second time to determine the behavior of the wall under repeated cyclic loading. This process was repeated until the specimen experienced a significant drop in strength. The specimen was not repaired in anyway in between each fatigue loading.

The specimens connected to the framing with 8 d common nails were much stronger and more ductile than any previously reported studies and as a result the capacity of the testing facility was reached before the specimens experienced a 20% drop in peak strength. Specimens A1, B, and C were placed under the first 37 cycles of the CUREE cyclic loading protocol and reached peak loads ranging from 73,840 N to 88,964 N and displacements ranging from 107 mm to 132 mm. Even though the specimens did not experience a drop in peak load, at the end of the initial 37 cycles their drift ratios ranged from 4.4% to 5.4%, which is far beyond the maximum allowable 2.5% drift ratio, stated in ASCE 7-05 [29].

A series of trend lines were used to estimate the failure point (where load resistance drops to 80% of the peak load) of the fatigued specimens. The trend lines used were developed by fitting third and fourth power polynomial curves to the data points already obtained from the testing. The curves made it possible to extend the test data points and predict the failure point. An average of the third and fourth polynomial curves was also plotted. In addition to that, an average of the failure curves of Specimens A3 and A4 was calculated and also plotted on the envelope curve graph. This was determined by finding the percentage that the displacement increased, beyond Δpeak, when Specimens A3 and A4 experienced a 20% drop in peak load. The hysteresis loops for Specimen C did not begin to level off towards the end of the test as they did for Specimens A1 and B. As a result, trend lines were drawn from both the actual peak point of the 37th cycle of the test (similar to Specimens A1 and B) and additionally from a point extended to a load increased by 15%.

The average of the percentage increase in displacement was calculated and that average percentage was applied to the Δpeak of the specimen to determine Δu. All of these trend lines and estimated failure points were then analyzed according to ASTM E2126 [27,28] and ICC-ES AC130 [4] and

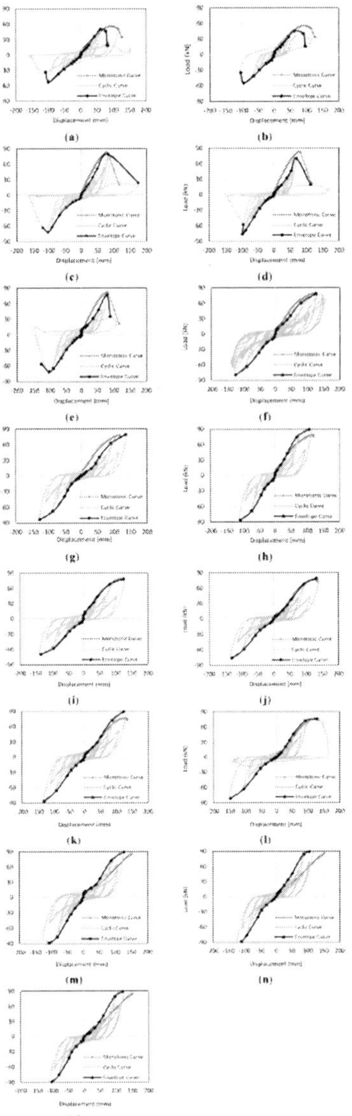

Figure 5. Load vs. displacement of SIP specimens under cyclic and monotonic loading. (a) Specimen A3-1C; (b) Specimen A3-2C; (c) Specimen A4-1C; (d) Specimen A4-2C; (e) Specimen A4-3C; (f) Specimen A1-1C; (g) Specimen A1-2C; (h) Specimen A1 Bearing-3C; (i) Specimen A1 Internal-4C; (j) Specimen B-1C; (k) Specimen B-2C; (l) Specimen B-3C; (m) Specimen C-1C; (n) Specimen C-2C; (o) Specimen C-3C.

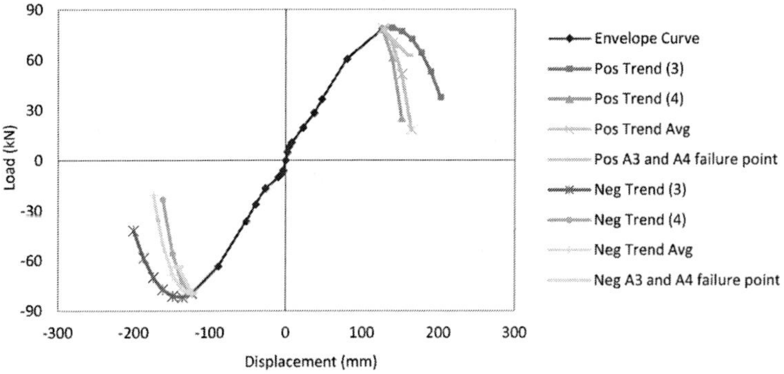

(a)

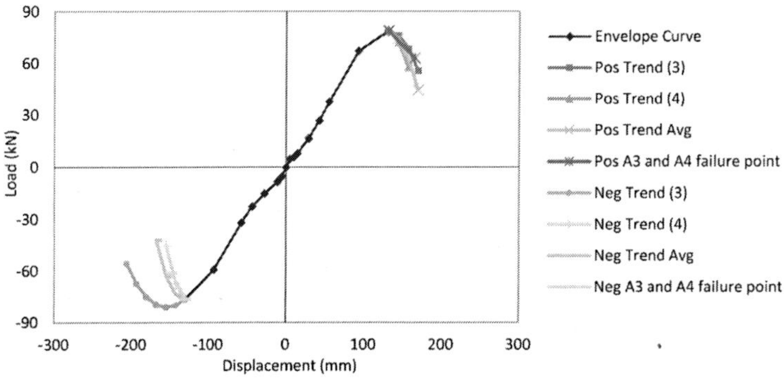

(b)

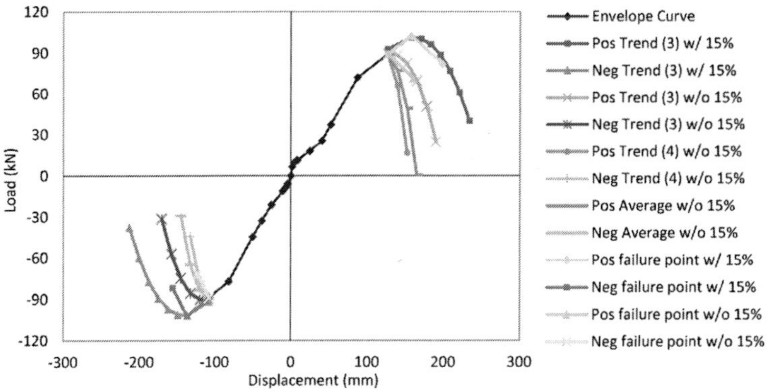

(c)

Figure 6. Envelope curve and trend lines developed to predict failure points. (a) Specimen A1-1C Trend Line Curves; (b) Specimen B-1C Trend Line Curves; (c) Specimen C-1C Trend Line Curves.

used to determine the minimum and maximum values of load, displacement, strength, stiffness, and shear modulus needed to describe the specimen's performance. Figure 5 shows the load vs. displacement graphs of the specimens under monotonic and cyclic loading along with the corresponding envelope curve. Figure 6 shows a representative envelope curve for specimens A1, B and C and the trend lines used to analyze their failure points. Refer to Terentiuk [1] for a more in depth description of the analysis.

On average, Specimen C had the least amount of loss or change after the first fatigue loading. The shear strength did not change at all and the shear modulus and ductility had less than a 10% decrease in value. The Fatigue 2 test did not have as much of an impact on the specimens as the Fatigue 1 test. Specimen A1 had the least amount of change after Fatigue 2 in comparison to Specimens C and B. Unlike Specimen B, which completely failed after the second fatigue, Specimen A1 was able to withstand a third fatigue test. Specimen C would have been able to withstand a third fatigue test as well but due to the insignificant change between the first two fatigues and the original cyclic loading the specimen was not fatigued for a third time. It was assumed that a third fatigue would produce similar results. Table 3 and Table 4 show the average percentage change in the characteristic values of each specimen under each fatigue loading.

It is important to note that the wood frame walls were sheathed on both sides making them much stronger than a wall with sheathing on one side only. Previous published research testing light-frame wood walls with single sided sheathing obtained significantly lower peak loads and displacements [18,19]. It can be assumed that the Specimens A1 and B would have outperformed the traditional wood frame specimen had it only been sheathed on one side.

Table 3. Average percentage loss (−) or gain (+) in characteristic values at strength limit state after fatigue tests of specimens.

Response Parameter	Fatigue 1			Fatigue 2			Fatigue 3
Specimens	C	A1	B	C	A1	B [1]	A1
Displacement	+13	+3	+10	0	−2	+3	+1
Shear Force	0	−18	−12	−9	−5	−4	−12
Shear Modulus	−7	−23	−19	−10	−4	−8	−13
Shear Strength	0	−17	−13	−9	−3	−4	−4
Elastic Shear Stiffness	−40	−53	−46	−15	−9	−10	−13
Ductility	−2	−8	−3	−2	+1	0	−7

[1] Specimen B-2C failed before a second fatigue test so the values in this column are not an average, they are the loss or gain experienced only by Specimen B-3C after Fatigue 2 test.

Table 4. Average percentage loss (−) or gain (+) in characteristic values at yield limit state after fatigue tests of specimens.

Response Parameter	Fatigue 1			Fatigue 2			Fatigue 3
Specimens	C	A1	B	C	A1	B [1]	A1
Displacement	+9	+11	+3	+5	−3	+5	+1
Shear Force	−9	−28	−21	−9	−5	−4	−12
Shear Modulus	−20	−36	−24	−12	−5	9	−13

[1] Specimen B-2C failed before a second fatigue test so the values in this column are not an average, they are the loss or gain experienced only by Specimen B-3C after Fatigue 2 test.

1 Specimen B-2C failed before a second fatigue test so the values in this column are not an average, they are the loss or gain experienced only by Specimen B-3C after Fatigue 2 test.

5.3. Evaluation by ASTM E2126 and Criteria

The following sections describe the analytical evaluation associated with each specimen based on the ASTM E2126 criteria [27,28]. In this paper, to determine the average response parameters the positive and negative envelope curves were analyzed individually and then the two values obtained were averaged, as described in ASTM E2126-08 [27]. According to the most recent version of ASTM E2126-11 [28], these parameters may be non-conservative when a specimen responds asymmetrically to the testing. Refer to Terentiuk [1] for the results found by analyzing the positive and negative envelope curves individually without averaging. Both the SIP specimens and the wood frame specimens were analyzed using the method described in ASTM E2126-08 [27]. The consistency in the analysis method should result in a fair comparison between the two types of specimens. The data obtained during the monotonic and cyclic loading of the wall systems was used to determine performance parameters of the various SIP wall designs. The shear strength was found by determining the absolute value of the load per unit length of the specimen, $V_{peak} = \frac{P_{peak}}{L}$. The secant shear modulus at both 0.4 Ppeak and Ppeak was found by using the relation: $G' = \frac{P}{\Delta} \times \frac{H}{L}$. The $\frac{H}{L}$ refers to the aspect ratio of the specimen. The cyclic ductility ratio is defined as the ratio of the ultimate displacement and the yield displacement, $D = \frac{\Delta_u}{\Delta_{yield}}$. An equivalent energy elastic-plastic (EEEP) curve was developed by circumscribing the area enclosed by the envelope curve. The enclosed area was bordered by the origin, the ultimate displacement, and the displacement axis of the envelope curve. The envelope curve consisted of the extreme points of the load-displacement hysteresis loops. An EEEP curve can be used as a visual comparison between differing wall designs and materials [27,28].

The calculations for the specimens that used trend lines were slightly different in that each step was repeated for each trend line. For instance, the calculations in Sections 9.1.1 through 9.1.4 of ASTM E 2126 [27,28] were followed to determine the performance parameters of Specimen A1 in terms of the third power polynomial trend line. Next, the fourth power polynomial trend line was analyzed in the same manner. This was repeated until all of the trend lines were analyzed. By determining the performance values according to each trend line, a range of values was determined for the specimen.

Table 5 compares the average characteristic values obtained by analyzing data from the cyclic tests performed on each specimen. The values in the table demonstrate the effect hardware and spline design have on the engineering values of a specimen. Specimens A3 and A4 were able to withstand the least amount of displacement and shear force before they failed. The peak displacement of Specimen A1 was slightly less than Specimen B which can be attributed to the difference in spline designs. The double 38 mm × 89 mm spline in Specimen B slightly reduced the shear modulus, shear strength, and elastic shear stiffness in comparison to the OSB spline in Specimen A1. Specimen C had the highest shear force, shear modulus, shear strength, and elastic shear stiffness. It is important to remember that the values found for Specimen C are based on a timber wall with sheathing on both sides, whereas actual construction methods typically have OSB sheathing on one side. Specimen A4 had the greatest shear modulus, it was about 13% greater than Specimen C and 23% greater than Specimen A3. Specimens A1 and B had the lowest shear modulus values.

As expected, fastener hardware had an effect on ductility. Specimen A1 had the highest ductility, followed by Specimens C and B. A specimen with high ductility has the ability to yield and deform inelastically without experiencing a significant loss of load resistance. However, the ductility property should be examined in conjunction with other characteristic values because a high ductility factor does not directly mean that the specimen will perform well under seismic loading. Specimen C had the highest shear strength out of all the specimens while Specimen A3 had the lowest shear strength. Specimen A4 was slightly lower than Specimens A1 and B, which have very similar shear strength values. Specimens C and A4 had very similar elastic stiffness values which was unexpected. The similar

Table 5. Average characteristic values of each specimen under cyclic loading.

Response Parameter	A3	A4	A1		A1 Bearing		A1 Internal		B		C	
			Min	Max	Min	Max	Min	Max	Min	Max	Min	Max
Δ_{max} (mm)	79.50	88.39	128.27	134.62	106.68		129.79		131.57	139.95	111.76	149.61
Δ_{yield} (mm)	69.34	76.71	108.71	111.00	87.88	90.17	94.49	95.76	114.05	133.10	96.52	128.52
F_{max} (N)	50,768	74,236	80,055	80,842	87,652		73,854		81,300	82,679	88,866	104,164
F_{yield} (N)	48,236	63,970	75,433	77,132	84,391	86,477	69,561	70,317	73,436	84,623	81,091	101,206
G' (N/mm)	664	857	602	625	822		571		597	621	697	800
G'_{yield} (N/mm)	708	847	699	700	979		800		653	655	785	850
Ductility	1.36	1.34	1.36	1.73	1.36	1.56	1.63	1.99	1.26	1.41	1.33	1.53
V_{peak} (N/m)	20,826	30,443	32,836	33,157	35,945		30,282		33,347	33,902	36,441	42,716
K_e (N/mm)	708	765	699	700	979		800		653	655	768	800

Note: $G' = (P/\Delta) \times (H/L)$. H = height of the wall, L = length of the wall; Ductility = Δ_u/Δ_{yield}; $V_{peak} = P_{max}/L$; $K_e = 0.4\ P_{peak}/\Delta_e$. Δ_e is Δ at the top edge of the wall at the corresponding $0.4\ P_{peak}$.

elastic stiffness values signify that both specimens will have reduced lateral drift during seismic loading, which will reduce nonstructural damage [18].

5.4. Allowable Drift Capacity

The allowable drift for service wind loading is found by using H/400. In this case, $\frac{H}{400} = \frac{2.4\text{ m}}{400} = 0.006\text{ m} = 6\text{ mm}$. The allowable seismic drift is 2.5% of the height of the specimen, or 0.025 × H = 0.025 × 2.4m = 0.061 m = 61 mm [2,3], or almost ten times the allowable wind drift. All of the SIP specimens performed much better under monotonic and cyclic loading than expected. Most A1, B, and C type specimens were able to withstand drifts of 127 mm (5.2% drift ratio) and greater, which is well beyond the practical application of these systems. Therefore, most of the specimens showed capacities at least twice the allowable seismic drift limit. This shows that the wall systems are not only strong but they are also ductile.

Parameters such as elastic stiffness, strength, and ductility are some of the factors which govern the response a shear wall has under seismic loading. These deformational characteristics are based on the wall's load-displacement relationship under cyclic loading. Specimen C was able to withstand the greatest force and displacement before the capacity of the wall began to decline. A larger peak load results in a larger load at the yield limit strength. As a result, Specimen C performed elastically under a higher load and displacement compared to the SIP specimens tested. Refer to Table 6 for the average peak load and displacement values of the specimens at their strength limit state.

The SIP specimens tested under monotonic loading had higher load capacities at both the allowable wind and seismic drifts than the specimens tested under cyclic loading. For Specimens A3, A4, A1 and B the load capacity at the allowable drift for service wind loading ranged from 35% to 50% greater during monotonic loading in comparison to cyclic loading.

In terms of cyclic loading, Specimen A1Internal-4C required the greatest load (10,208 N) to be pushed to a drift of 6 mm. Specimen A1Bearing-3C was high as well, within 10% of Specimen A1Internal-4C. Specimens A3, A4, and C had load capacities ranging from about 5631 N to 8990 N, which were 55% to 88% of the load required to displace Specimen A1Internal-4C. The largest amount of force was needed to push Specimen A1Bearing-3C to the allowable seismic drift of 61 mm out of all the specimens. Specimen C and A4 needed about 20% less force than A1Bearing-3C.

Table 6. Maximum measured load and corresponding drift, and load capacity corresponding to allowable drift under wind and seismic loading condition.

Specimen	Maximum Load Measured and Corresponding Drift		Capacity N at 6 mm Drift (Allowable drift for wind)	Capacity N at 61 mm Drift (Allowable drift for seismic)	V_{peak} (N/m)	Allowable Load (V_{peak}/F.S. [1])
	Max. Load (N)	Drift (mm)				
A3 Monotonic	55,698	92	11,707	45,877	22,838	7,618
A3-1C Cyclic	51,494	80	6,512	41,620	21,116	7,034
A3-2C Cyclic	50,036	79	8,389	41,905	20,518	6,844
A4 Monotonic	82,791	79	12,112	74,193	33,958	11,324
A4-1C Cyclic	75,420	93	7,584	47,660	30,937	10,317
A4-2C Cyclic	73,276	83	5,631	49,773	30,047	10,011
A4-3C Cyclic	74,006	89	8,678	50,543	30,353	10,113
A1 Monotonic	78,133	114	11,863	57,522	32,046 to 32,163	10,682 to 10,721
A1-1C Cyclic	78,867	125	8,887	44,738	32,338 to 32,980	10,784 to 10,989
A1-2C Cyclic	81,243	131	2,882	43,839	33,316	11,105
A1 Brg-3C Cyclic	87,648	107	9,163	61,631	35,945	11,981
A1 Int-4C Cyclic	73,855	130	10,208	45,352	30,280	10,098
B Monotonic	76,466	131	7,811	45,814	31,360	10,449
B-1C Cyclic	77,443	131	5,351	38,159	31,754 to 32,732	10,580 to 10,916
B-2C Cyclic	87,638	127	7,063	45,370	35,943 to 36,483	11,981 to 12,156
B-3C Cyclic	78,810	137	2,851	40,806	32,323 to 32,499	10,770 to 10,828
C Monotonic	90,539	178	3,799	32,517	37,127 to 46,277	12,376 to 15,426
C-1C Cyclic	88,875	117	8,990	50,847	36,441 to 41,914	12,147 to 13,971
C-2C Cyclic	88,777	107	5,743	50,407	36,412 to 44,074	12,137 to 14,691
C-3C Cyclic	88,942	111	6,294	49,255	36,470 to 42,162	12,157 to 14,054

[1] Factor of Safety (F.S.) = 3.0 [3].

5.5. Energy Dissipation

The energy dissipation of a specimen is determined by finding the area enclosed by the hysteresis loops obtained from the load vs. displacement graph of a specimen under cyclic loading. In this paper, the trapezoid rule was used to determine the area within the hysteresis loops. To perform well during an earthquake, a structure must be able to dissipate large amounts of energy. When a shear wall is within its elastic limit, it will not dissipate any hysteretic energy. Figure 7 and Figure 8 show graphs of the average energy dissipated per cycle and the average total (cumulative) energy dissipated up to the current cycle of the specimens tested under cyclic loading. The energy dissipated in the early cycles is minimal compared to the large spikes found in the later primary cycles. When a shear wall is pushed past its elastic limit, the energy is dissipated through inelastic behavior or fracture of fasteners/connection materials. A minimal amount of energy is also dissipated through the friction forces created by panel sheathing rubbing up against an adjacent panel or framing members

[30]. Specimens A1Bearing-3C and A1Internal-4C were not included in their appropriate specimen averages because they were not an identical replica of the original walls tested.

Specimens A1 and C had the ability to dissipate the largest amount of energy within 37 cycles, while Specimen A4 dissipated the least amount of energy within the same number of cycles. This is consistent with the strength and displacement capacities of the specimens. Specimen A4 was able to withstand three more cycles of the CUREE loading protocol than Specimen A3, and at the forty-first cycle, Specimen A4 dissipated 24% more cumulative energy than Specimen A3.

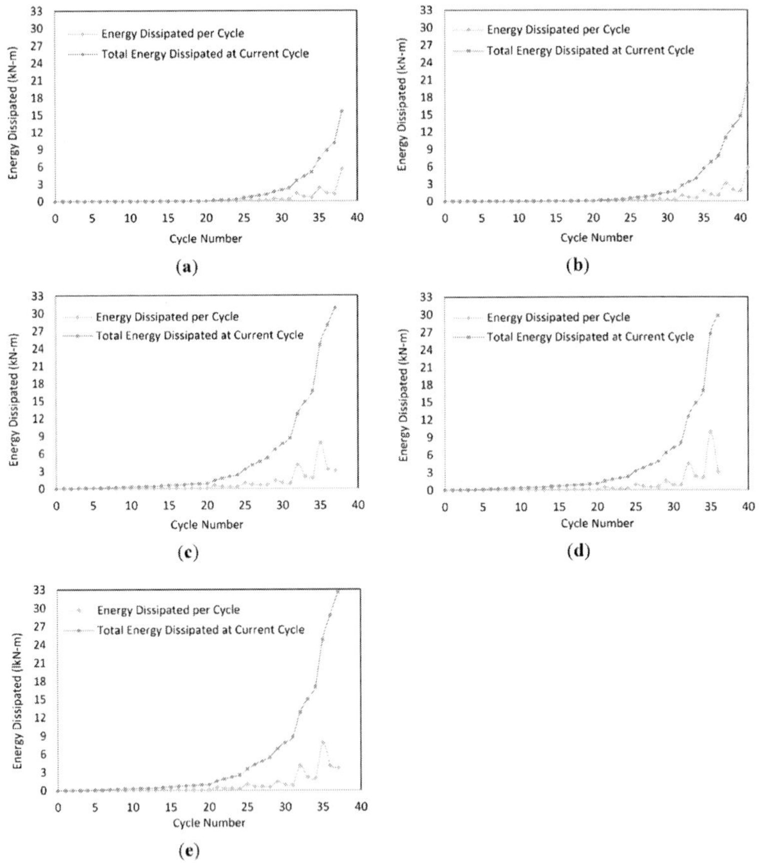

Figure 7. Average energy dissipation of SIP specimens under cyclic loading. (a) Specimen A3; (b) Specimen A4; (c) Specimen A1; (d) Specimen B; (e) Specimen C.

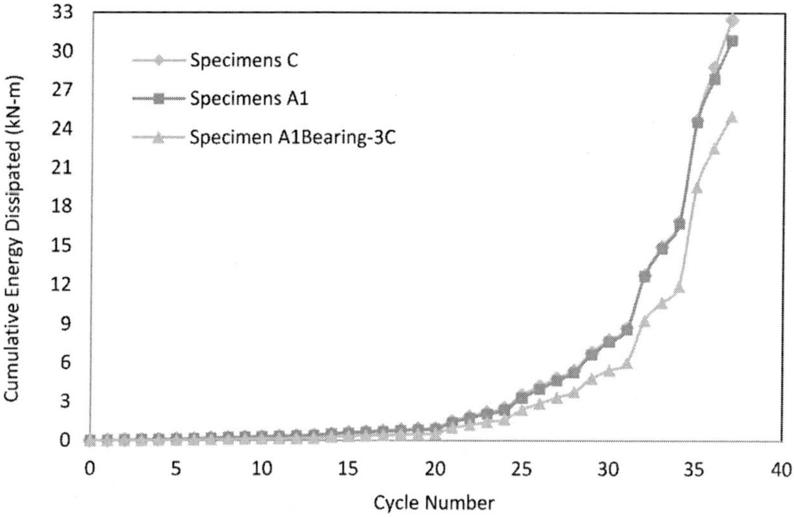

Figure 8. Comparison between cumulative energy dissipation of wood frame wall and SIP specimens.

6. COMPATIBILITY WITH WOOD FRAME SHEAR WALL—ICC-ES AC130, ICC-ES AC04 APPENDIX A, AND NTA IM 14 TIP 10.0

ICC-ES AC130 [4] was followed to determine if the specimens were deemed seismically compatible to a code-defined seismic-force resisting system. First, the ultimate displacement of the specimen divided by the displacement at the ASD design load must be greater than or equal to 11, or $\frac{\Delta_u}{\Delta_{ASD}} \geq 11$. The ASD design load is 70% of the load of the specimen found at a displacement of 15 mm. Next, the ultimate displacement has to be greater than 2.8% of the height of the specimen, as described in the equation, $\Delta u \geq 0.028H$. Finally, the ratio of the peak load to the ASD design load must be between or equal to 2.5 to 5.0, $2.5 \leq \frac{P_{peak}}{P_{ASD}} \leq 5.0$. If all of these requirements are met, the prefabricated panels should be deemed usable as a seismic force resisting system and the specimen can be assigned the following IBC values:

(1) Response Modification Coefficient: R = 6.5;

(2) System Overstrength Factor: Ω0 = 3;

(3) Deflection Amplification Factor: Cd = 4.

ICC-ES AC130 was developed to be used with prefabricated wood shear panels but was applied to structural insulated panels as well in this study. Specimens A1 (8 d common nails, OSB surface spline), B (8 d common nails, double 39 mm × 89 mm lumber spline), and C (wood-frame) met the requirements stated in ICC-ES AC130; therefore, they can be used within a seismic-force resisting system. However, the specimens did not meet Section 5.3.4 of ICC-ES AC130 [4] because the $\frac{P_{peak}}{P_{ASD}}$ ratios turned out to be greater than 5.0. In order to be considered compliant, the evaluation report for the panel must include "a requirement that collectors and their connections, bearing and anchorage of the panel, and the lateral load path to the panel are designed in accordance with the special load combinations of Section 12.4.3 of ASCE 7, using Em where Em is calculated using the test panel overstrength" [4]. Similar to Specimens A1, B, and C, the fatigue tests of these specimens were deemed seismically compatible but did not meet the full criteria of Section 5.3.4 in ICC-ES AC130 [4].

Unlike ICC-ES AC130, Appendix A of ICC ES AC04 [3] applies directly to SIP sandwich panels under cyclic loading. The ductility, drift, and overstrength compatibility requirements are exactly the same as ICC-ES AC130 [4]. Axial load was not applied to the specimens tested in this study so they would be considered non-load bearing Assembly B (Specimen B, lumber spline) and Assembly C (Specimen A, surface spline) SIP shear walls.

Instead of comparing test values of a SIP specimen to a predetermined performance criterion as seen in ICC-ES AC130 [4] and ICC ES AC04 [3], the previous Appendix A of the 2005 version of ICC-ES AC04 [2] allowed SIPs to be used in all of the IBC seismic design categories if it was shown to be equivalent to a light-framed wood-based shear wall under cyclic loading. The following requirements had to be met: the peak strength of the SIP specimen had to be within 90% of the benchmark (wood-based shear wall), the stiffness of the SIP specimen had to be within 85% of the benchmark, and the load capacity of the panel at the allowable story drift under seismic loading (Δall = 61 mm) had to be within 85% of that for the wood-based shear wall. In addition to these three requirements, the cumulative energy dissipated by the SIP specimen had to be within 85% of that of the benchmark. In this study, Specimen C was considered the benchmark and after reviewing the peak strength (Table 5), stiffness, allowable story drift (Table 6), and the cumulative energy dissipated (Figure 7 and Figure 8) of the six SIP specimen designs, Specimens A1 (8 d common nails, OSB surface spline) met all of the requirements. Therefore, according to Appendix A of ICC-ES AC04 [2], Specimen A1 was deemed equivalent to a wood-frame wall. The implication is that this system may be permitted to be used as shear walls in buildings located in Seismic Design Categories A through F.

NTA IM 14 TIP 10.0 [5] has similarities to both Appendix A of ICC-ES AC04 [2] and ICC-ES AC130 [4]. Similarly to ICC-ES AC04 [2], the performance of a SIP specimen under cyclic loading is compared to the performance of a conventional wood frame wall. Unlike ICC-ES AC04, according to NTA IM 14 TIP 10.0, if the SIP demonstrates equivalence to the wood frame benchmark assembly, it can be assigned the strength and seismic design parameters of the benchmark assembly. More specifically, it should be deemed equivalent to System A13 in ASCE 7 Table 12.2-1, R = 6.5, Ωo = 3, and Cd = 4 [6]. These are the same seismic factors applied in ICC-ES AC130 [4]. In this study, Specimen C was used as the benchmark for comparison with the structural insulated panels. The difference between ICC-ES AC130 and NTA IM I4 TIP 10.0 is that the Allowable Stress Design (ASD) load used in the equivalency analysis is based on IBC Table 2306.4.1 [20] or a code research report instead of Section 5.1.3 of ICC-ES AC130 [4].

To meet the first performance requirement of Section 7 in the NTA IM 14 TIP 10.0 [5], the peak strength load of the SIP panel cannot be less than 90% of that of the wood frame shear wall (Specimen C). Next, the displacement at the ASD design load, P = 9,963.5 N [20], for the SIP panel cannot be less than 85% of that of the benchmark (Specimen C). The ultimate displacement (Δu) shall not be less than 85% of the benchmark specimen. The ratio of the ultimate displacement to the ASD design load displacement shall not be less than 85% of that of the benchmark specimen. The load at the maximum allowable story drift (Δall = 61 mm) cannot be less than 85% of that of Specimen C. The final requirement states that the cumulative energy dissipated by the SIP specimen cannot be less than 85% of that for Specimen C.

With reference to Table 7, according to NTA IM 14 TIP 10.0 [5], Specimen A1 is equivalent to a wood frame wall under cyclic loading. As a result, the following seismic factors may be applied: R = 6.5, Ω0 = 3, and Cd = 4.

Table 7. Data to Meet Performance Requirements of NTA, Inc. [5].

Specimen	P_{peak} (N)	P_{peak}/P_{ASD}	Δ_u (mm)	Δ_{ASD} (mm)	Δ_u/Δ_{ASD}
C-1C	88,875	8.92	129.29	7.11	18.18
C-2C	88,777	8.91	122.68	13.46	9.11
C-3C	88,942	8.93	127.00	10.16	12.5
Average C	88,865	8.92	126.32	10.24	13.26
A1-1C	78,867	7.92	140.72	8.13	17.31
A1-2C	81,243	8.15	158.24	20.83	7.60
Average A1	80,055	8.04	149.48	14.48	12.46

7. SUMMARY, CONCLUSIONS, RECOMMENDATIONS, AND LIMITATIONS

In this study, a total of 21 wall specimens were tested under monotonic and cyclic loading. Characteristic values such as shear modulus and shear strength were found for each specimen, as well as allowable drift capacity and energy dissipation during cyclic loading. Parameters such as fastener hardware, spline design, hold-down anchor location, and sheathing bearing were adjusted on 2.4 m × 2.4 m structural insulated panels to determine their effect on the performance of the shear walls under monotonic and cyclic loading. The SIP specimens were compared to a traditional wood frame wall (but with sheathing on both sides) under identical loading procedures.

The mode of failure for all of the specimens occurred either in the fastener hardware or the OSB sheathing. The SIP specimens (A1, A3, A4) with the OSB surface splines typically had failure in the sheathing along the vertical spline connection. The staples withdrew and sheared, the screws sheared and the nails withdrew and caused sheathing tear-out. Specimen B failed when the two 38 mm × 89 mm spline members separated, allowing the SIP panels to rotate independently of each other. Additional 16 d common nails at a reduced spacing would significantly increase the capacity of the specimen. The specimens with nails withstood a peak load 37% higher and a peak displacement 38% higher than the stapled specimens. The nailed specimens also withstood a peak load 7% higher and a peak displacement 31% larger than the screwed specimens. The specimens held together with screws had a sudden and brittle failure due to the shear failure of the fasteners. The ductile nature of the 8 d common nails allowed Specimens C and A1 to dissipate a greater amount of energy than the screwed and stapled specimens.

In order to determine the effect design elements had on the specimen's performance, the fastener hardware was held constant (8 d common nails) and the spline design, hold-down anchor location, and sheathing bearing were adjusted. The spline design did not have a significant effect on the performance of the SIP specimens. The load, displacement, and ductility of Specimen A1 (8 d common nails, surface spline) and Specimen B (8 d common nails, double 38 mm × 89 mm lumber spline) were within 10% of each other. By placing the hold-down anchor on the interior of the SIP wall in rectangular cut-outs (Specimen A1 Internal), the specimen's ductility and elastic shear stiffness increased by about 13%. The specimen also required the greatest amount of force out of all of the SIP specimens to reach the allowable wind drift (6 mm). In actual field conditions, SIP sheathing often bears directly on the sill plate. During testing, this caused more extensive damage to the panels during cyclic loading than the specimens which lacked sheathing bearing. There was also a moderate increase in peak load (8%) and peak displacement (17%) in Specimen A1 Bearing. The elastic shear stiffness also increased by 29%.

Similar to results in previous publications [24], monotonic loading produced non-conservative results in comparison to cyclic loading. The peak load capacities at the allowable wind drift of 6 mm for the specimens averaged 35% to 50% greater under static loading versus cyclic loading. This is consistent with Section X2.2.2.2 in ASTM E2126 [28] that suggests that the reference deformation in the CUREE loading protocol be 60% of the monotonic deformation capacity. This is because the cumulative damage caused under cyclic loading leads to a quicker reduction in strength than damage incurred under monotonic loading.

Specimens that did not fail after the first 37 cycles of the CUREE cyclic loading protocol were loaded a second, third, and fourth time if possible to determine their ability to withstand repeated cyclic loading. Specimen C was able to retain the greatest amount of strength after Fatigue 1 test in comparison to Specimens A1 and B. Specimen A1 had the smallest decrease in strength after Fatigue 2 test. During fatigue loading of Specimens C, A1 and B the elastic shear stiffness experienced the largest reduction out of all the structural properties calculated. The specimens had an average loss of 40% to 53% after Fatigue 1 test and an average loss of 9% to 15% after Fatigue 2 test.

The results found from monotonic and cyclic testing were examined under three different seismic evaluation procedures, ICC-ES AC130 [4], ICC-ES AC04 [2,3], and NTA 1M 14 TIP 10.0 [5]. The SIP specimens connected with

8d common nails and either the surface splines or a double 38 mm × 89 mm lumber splines (Specimens A1, A1 Bearing, A1 Internal, and B) met the requirements stated in ICC-ES AC130 [4] and ICC-ES AC04 Appendix A [3], which allows them to be used within a seismic force resisting system with the following values: R = 6.5, Ωo = 3, and Cd = 4. Specimen A1 was the only SIP design which met the additional requirements stated in ICC-ES AC04 [2] and NTA 1M 14 TIP 10.0 [5]. As a result and based on such criteria, Specimen A1 may be permitted to be used as shear walls in buildings located in Seismic Design Categories A through F with the following seismic values: R = 6.5, Ωo = 3, and Cd = 4. Out of the various SIP specimen designs, A1 was the most effective design in terms of: load capacity, ductility, resistance under fatigue loading and seismic compatibility.

This research should be considered as preliminary testing, which can be used to provide a better understanding of the performance of structural insulated panels with varying parameters under monotonic and cyclic loading. Each specimen was tested once under monotonic loading and the minimal requirements for cyclic testing according to the ASTM standards. To provide a more thorough investigation, additional monotonic and cyclic testing should be performed. The specimens in this study proved to be much stronger than stated in previously published research. A testing facility with a load capacity of at least 120,096 N and a minimal drift capacity of 254 mm is required to bring the specimens to their ultimate failure should future testing be performed. ASTM 2126 [27,28], which was followed in this study, limits the amount of axial loading applied to the specimens under lateral loading. Future testing should place the SIPs under biaxial loading in order to mimic actual field conditions and demonstrate the panel's ability as a load-bearing wall under cyclic loading.

ACKNOWLEDGEMENTS

This research presented in this report was made possible through partial support by the Structural Insulated Panel Association (SIPA). The SIP panels used for testing were donated by Timberline Panel Company. These contributions are gratefully acknowledged. In development of the research program and throughout the study the researchers benefitted from suggestions and comments by several individuals including Bill Wachtler, Jim DeStefano, Scott Maxwell, Jim Whalen, Borjen Yeh, Todd Bergstrom, Eric Tompos, Joe Hagerman, and Joe Pasma. For the experimental study the help of the following students is acknowledged: Dan Clark, Dan Navarrete, Joe Ridgeway, and Hiroki Ota. In particular, the invaluable help of BERL

Laboratory Supervisor, Paul Kremer, in all aspects of the experimental work including test setup design, data acquisition system design, and data analysis is acknowledged. The views and opinions expressed in this report are those of the authors and do not necessarily represent SIPA's position.

AUTHOR CONTRIBUTIONS

The first author Stefanie Terentiuk was the graduate student working on the research project that led to her M.S. thesis and a research report. The second author Ali Memari was the project PI and thesis advisor. Both authors worked on the manuscript through the final submitted form.

REFERENCES

1. Terentiuk, S. Parametric Study of Structural Insulated Panels Under Monotonic and Cyclic Loading. Master's Thesis, The Pennsylvania State University, University Park, PA, USA, May 2009.
2. ICC-ES AC04. In Acceptance Criteria for Sandwich Panels; ICC Evaluation Service Inc.: Whittier, CA, USA, 2005.
3. ICC-ES AC04. In Acceptance Criteria for Sandwich Panels; ICC Evaluation Service Inc.: Whittier, CA, USA, 2012.
4. ICC-ES AC130. In Acceptance Criteria for Prefabricated Wood Shear Panel; ICC Evaluation Service Inc.: Whittier, CA, USA, 2012.
5. NTA IM 14 TIP 10.0. In Qualification of Wood Structural Panel Faced SIPs for Use in Seismic Zone D, E and F; NTA, Inc.: Nappanee, IN, USA, 2014.
6. NTA IM 14.1. In General Overview of NTA Structural Insulated Panel Qualification and Quality Assessment Procedures; NTA, Inc.: Nappanee, IN, USA, 2008.
7. Jamison, J.B. Monotonic and Cyclic Performance of Structurally Insulated Panel Shear Walls. Master's Thesis, Virginia Polytechnic Institute and State University, Blacksburg, VA, USA, 10 December 1997. [Google Scholar]
8. Mosalam, K.M.; G&#xFCnay, M.S. Seismic Performance of Energy Efficient Structural Insulated Panels. Wood Des. Focus 2012, 22, 12–18.

9. Keith, E.L. Standardization Testing of Structural Insulated Panels (SIPs) for the Structural Insulated Panel Association; APA The Engineered Wood Association: Tacoma, WA, USA, 2006.

10. Kermani, A.; Hairstans, R. Racking Performance of Structural Insulated Panels. J. Struct. Eng. 2006, 132, 1806–1812.

11. Filiatrault, A.; Foschi, R.O. Static and Dynamic Tests of Timber Shear Walls Fastened with Nails and Wood Adhesive. Can. J. Civil Eng. 2001, 18, 749–755.

12. ICC Evaluation Service, Inc. ICC-ES Legacy Report for INSULSPAN Structural Insulated Panels; ICC Evaluation Service, Inc.: Whittier, CA, USA, 2004.

13. INSULSPAN. Technical Bulletin for Racking Shear Strength Test Results; INSULSPAN Structural Insulating Panel System: Blissfield, MI, USA, 2007.

14. ICC Evaluation Service, Inc. ICC-ES Legacy Report for Intermountain Building Panels L.L.C.; ICC Evaluation Service, Inc.: Whittier, CA, USA, 2003.

15. R-Control Building Systems. Available online: http://www.r-control.com/downloads/techbulletin/rcontrol2067.pdf (accessed on 29 April 2008).

16. Structural Engineers Association of Southern California (SEAOSC). Standard Method of Cyclic (Reversed) Load Test for Shear Resistance of Framed Walls for Buildings; Structural Engineers Association of Southern California: Whittier, CA, USA, 1996.

17. Architectural Testing. Performance Test Report Rendered to: Agriboard Industries Report; Architectural Testing: York, PA, USA, 2005.

18. Johnston, A.R.; Dean, P.K.; Shenton, H.W., III. Effects of Vertical Load and Hold-Down Anchors on the Cyclic Response of Wood Framed Shear Walls. J. Struct. Eng. 2006, 132, 1426–1434. [Google Scholar] [CrossRef]

19. Lebeda, D.J.; Gupta, R.; Rosowsky, D.V.; Dolan, J.D. Effect of Hold-Down Misplacement on Strength and Stiffness of Wood Shear Walls. Pract. Period. Struct. Des. Constr. 2005, 10, 79–87.

20. International Code Council, Inc. 2006 International Building Code; International Code Council, Inc: Country Club Hills, IL, USA, 2006.

21. The Canadian Construction Materials Center. Available online: http://irc.nrccnrc.gc.ca/pubs/ci/v2no1/v2no1_7_e.html (accessed on 28 February 2009).

22. ASTM Standard E564-06. In Standard Practice for Static Load Test for Shear Resistance of Framed Walls for Buildings; American Society for Testing and Materials: West Conshohocken, PA, USA, 2006.

23. Krawinkler, H.; Parisi, F.; Ibarra, L.; Ayoub, A.; Medina, R. Development of a Testing Protocol for Wood Frame Structures; Stanford University: Stanford, CA, USA, 2001.

24. Gatto, K.; Uang, C. Effects of Loading Protocol on the Cyclic Response of Woodframe Shearwalls. J. Struct. Eng. 2003, 129, 1384–1393.

25. Carradine, D.M.; Woeste, F.E.; Dolan, J.D. imber Frame and Structural Insulated Panel Building Design Concepts. In Proceedings of the 8th World Conference on Timber Engineering, Lahti, Finland, 14–17 June 2004; pp. 1–4.

26. Terentiuk, S.; Memari, A. In-Plane Monotonic and Cyclic Racking Load Testing of Structural Insulated Panels. J. Arch. Eng. 2012, 18, 261–275.

27. ASTM Standard E2126–08. In Standard Test Methods for Cyclic (Reversed) Load Test for Shear Resistance of Vertical Elements of the Lateral Force Resisting Systems for Buildings; American Society of Testing and Materials: West Conshohocken, PA, USA, 2008.

28. ASTM Standard E2126–11. In Standard Test Methods for Cyclic (Reversed) Load Test for Shear Resistance of Vertical Elements of the Lateral Force Resisting Systems for Buildings; American Society of Testing and Materials: West Conshohocken, PA, USA, 2011.

29. SEI/ASCE 7-05. In Minimum Design Loads for Buildings and Other Structures; American Society of Civil Engineers: Reston, VA, USA, 2005.

30. Bredel, D.H. Performance Capabilities of Light-Frame Shear Walls Sheathed with Long OSB Panels. Master's Thesis, Virginia Polytechnic Institute and State University, Blacksburg, VA, USA, April 2003.

INDEX

H

M

O

P

R

S

T

W